The Yachtsman's Naturalist

by *Maldwin Drummond* and *Paul Rodhouse*
with Foreword by *Sir Peter Scott* CBE, DSC

Illustrated by John Rignall

Angus & Robertson · Publishers

Dedication

Eileen & Ken Rodhouse; Frederica, Annabella & Aldred Drummond; Sophie, Ariane & Laura Turner-Laing

Angus & Robertson · Publishers

London · Sydney · Melbourne · Singapore · Manila

First published in Great Britain by Angus & Robertson (UK) Ltd, 10 Earlham Street, London, in 1980

Copyright © Maldwin Drummond and Paul Rodhouse, 1980
Illustrations copyright © John Rignall, 1980
Figures 4, 5, 8–18, 22–33 © Maldwin Drummond, 1980

ISBN 0 207 95808 4

Typeset by Computacomp (UK) Ltd, Fort William, Scotland
Printed and bound in Great Britain by
R. J. Acford Ltd., Industrial Estate, Chichester, Sussex

Contents

LIST OF COLOUR PLATES

LIST OF FIGURES

LIST OF ILLUSTRATIONS
IN TAXONOMIC DESCRIPTIONS

ACKNOWLEDGEMENTS

We would particularly thank Sir Peter Scott for his Foreword. He is the foremost yachtsman/naturalist and his words may encourage others to take a closer look at the sea.

It is no exaggeration to say that this book is made by the colour illustrations. We are greatly indebted to John Rignall for the excellence of his work and the care and enthusiasm which are the hallmark of his efforts.

We owe a debt of gratitude to Dr John Mercer, Deputy Director of the Shellfish Research Laboratory, University College, Galway at Carna, Ireland, for his help and encouragement. Debbie Brosnan, now Rodhouse, and Gilly Drummond contributed ideas and constructive criticism.

The book would not have been published if it had not been for the enthusiastic leadership of Ian Dear of Angus & Robertson or the hard work of Rosalie Hendey and Una Curran who typed the manuscript. Our grateful thanks go to them.

The authors received help and encouragement from many quarters. We would like to thank the following for their part in making the book possible; Joe Bossanyi, Philip Bradburn, Wright and Pat Britton, Lord Camrose, David Crook, Sidney Crook, Bend'or Drummond, Agustin Edwards, Finn and Petra Engelsen, Robin Ford of Watkins & Doncaster, Arthur Frank—late of Charles Frank Griffin & George, Dr Francis Greaves, Eric Kay, R. D. Leakey of R. & B. Leakey, Sandra Leche, Cyril Lucas, David Miller, David Milln, Colin Mudie, Brian Ottway, Peter Plumb, Air Commodore Paddy Quinnell, the late George Smith, Peter Stanbury, Tony Whilde.

The firms, associations and societies given below were of great assistance too and we thank them.

Big Jon Inc. of Traverse City, Michigan, USA
International Paints
The Marine Biological Association of the United Kingdom, Plymouth
The Royal Botanic Gardens, Kew
The Royal Society for the Protection of Birds, Sandy, Bedfordshire
Simpson Lawrence Ltd, Glasgow

Foreword

Sir Peter Scott CBE, DSC

Our creeks and estuaries have, for ages, been the feeding grounds and roosting places for many of Europe's most important migratory birds. The salt marshes and mudflats are especially rich in the food they need, and attract huge flocks of waders, ducks and geese. Since the War, these once quiet and remote places have become valuable to man for another purpose—a harbour for small boats. Estuaries that once supported a few swinging moorings now boast of a marina cut into their banks. These days people enjoying a summer view from the beach may see more sail than in the days when the world's commerce relied on canvas alone. A revolution has occurred that seriously threatens the coastline of our country and places our wetlands in danger.

Sailing is one of the most valuable and valued of sports and pastimes, but those who take part in it must be careful to avoid spoiling the environment that is an integral part of their enjoyment.

I welcome this book because it will help yachtsmen to understand their environment, to recognise and value the other forms of life which share it with them and, in so doing, to enjoy both the sea and the shore all the more.

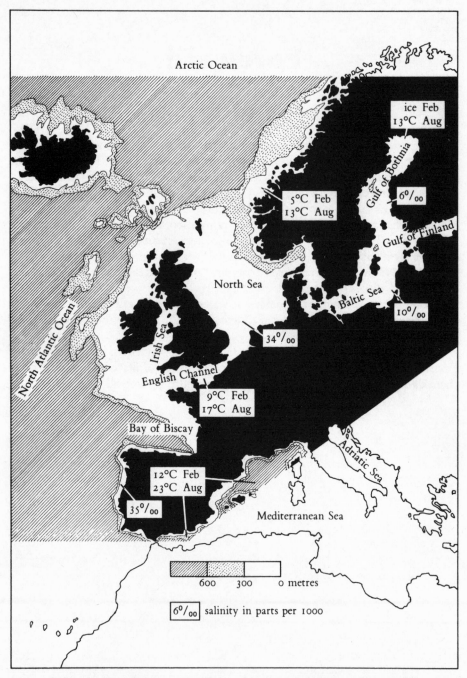

Map of the shore and seas covered by this book, showing the average water temperature and salinity.

Introduction

The sailing yacht has great advantages for the naturalist, for it creates less disturbance than almost any other form of transport and can be a comfortable roost for the observer, looking at estuary or rocky haven, miles from other distractions. A small boat seems at home in those surroundings. Few sailors seem to consider the creatures and plants that accompany them in the marine environment, whether on, below or above the surface, or living on the shores of the anchorages that they rely on for rest and recreation. There is a vast dimension of interest awaiting those who take even an elementary look at this natural world. This book is intended to provide an introduction to the ecology of the sea and shore of north-west European waters, and to supply the yachtsman/naturalist with an identification guide to most of the common and some of the more unusual plants and animals he is likely to meet while cruising at sea, at anchor, and ashore on the beach, mudflat or saltmarsh. It would clearly be impossible to cover every species in a book which includes so many different kinds of habitat, and so those examples given have been chosen as representative of the major taxonomic groups and for the part they play in the ecology of the sea and shore.

Part One covers the shore, the underwater world and the open sea. Each chapter is divided into two sections; the first section gives a descriptive account of the ecology of each major habitat, and the second gives descriptions of species for identification. Nearly all of the species described are illustrated, most in the colour plates and a number by black and white drawings. Colour illustrations are indicated by a reference in **bold** type, with the plate number in Roman numerals followed by the individual number in Arabic (e.g. **VIII, 7**). Black and white drawings will be found in the second section of Chapters One, Two and Three, and reference to them is by page number. (There are also line drawings of a more general nature, especially in the second part of the book, and these are identified by figure numbers.) Plants and animals have been arranged according to habitat as far as possible, but it should be remembered that although some species are very restricted in their range, others may be found in a variety of habitats. Where geographical distribution is given in the descriptive sections, this can only be taken to be where records exist for the species. When a species is not said to come from a particular area, it does not necessarily mean it is not present, but that it has not yet been recorded from there. Furthermore, plants and animals are not uniformly distributed, and the scarcity rating given is only a rough and ready guide. It should also be noted that the colours of many marine invertebrates and fish are extremely variable, and so this feature alone cannot be used as a sure means of identification. If an animal or plant cannot be identified with the aid of this book, the reader is advised to consult the relevant books given in the Bibliography.

Part Two is devoted to explaining how the sailor/naturalist may make the best use of his yacht. There are many excellent books on seamanship, but they tend to encourage the sailor away from the shore and inlets. These are the places the naturalist will wish to visit, and Chapter Four explains how to do this with safety and skill.

Chapter Five looks at the equipment that may be carried aboard a small yacht to help the sailor observe and measure the sea and shore environment. By simple observation and measurement, the world described in the first section comes alive. Discoveries can be made, and an added dimension given to small boat sailing.

The amateur yachtsman has the ability to go places denied to those who rely on other forms of transport. Remote islands have lost their human population because of the difficulties of modern servicing and the growing demands of the young for things only obtainable on the mainland. There is no way now of visiting many of them, except by hiring a boat and that is difficult and expensive. These islands are, therefore, mostly undisturbed and the sailor/naturalist has unique opportunities of studying small communities of plants and animals that live by the sea. With this privilege, of course, goes the need to respect such living laboratories, so that they are undamaged for others to examine, note and enjoy too.

An understanding, and a wish to know more, is a tenet of the serious sailor. Knowledge of his chosen environment gives depth and greater pleasure to his pursuit and enables him to add other skills to seamanship, navigation and small boat management, while at the same time conferring the privilege of passing such insight to his children and others. Such study and observation can increase enthusiasm for the sea. All this helps those who enjoy the water to understand the dangers of pollution. The destruction of a major habitat by vast accidents such as the *Amoco Cadiz* off Ushant or that of the *Torrey Canyon* off Britain's Scillies is all too obvious. In a lesser way, the thoughtless pumping of an oil-scarred bilge and even the disposal of an unwanted plastic bag into the sea create problems for wildlife. Waste, selfishly disposed of, has a habit of spoiling something somewhere, whether by upsetting the ecological balance of a tiny community, or just making the place that much more disagreeable for the next human being. The leisure of one surely must not spoil the leisure of another; the knowledge of how the natural communities live, by our own observation, can do more to give proper value to its protection than any amount of exhortation by others.

It is, therefore, particularly important in carrying out such observation not to damage the habitat under study. Plants and animals should not be removed except for close observation, preference being given to recording the life of, say, the rock pool or saltmarsh by using the sketch book or camera. Such material makes a valuable addition to the yacht's log or record of a visit to the shore, as explained in Chapter Five.

If specimens are removed to the ship's aquarium for close study and further photography, they should be returned undamaged to the place where they were found. The enthusiasm of the nineteenth-century naturalists and those who followed their example by collecting without heed to the injury they were inflicting, is a warning to us all.

The following pages encompass European and Scandinavian waters: the seas and coastline that stretch in variety from Portugal northward round Ireland, touching Shetland and on to north Norway. This imaginary line encloses within it the coast of continental Europe, the rich waters of the Irish and North Seas, the English Channel and the Baltic.

Before going into details of particular environments it is worth considering in general terms their composition, and the factors that influence the distribution of species.

THE MARINE ENVIRONMENT

The appearance of the seas of north-west Europe as we see them on the map today was not attained until the end of the last ice age. During the glacial periods there was considerable shifting of boundaries between land and sea, partly because the weight of the ice tended to depress the earth's crust and partly because of the changes in the amount of water in the oceans brought about by freezing and thawing of the ice. Until the end of the last ice age, much of the North and Irish seas were above water and the English Channel was non-existent. As the ice melted the sea level rose, flooding these areas, and the land, previously covered by ice, regained its isostatic equilibrium. As the ice retreated, species characteristic of warmer waters moved into our seas although a few relict Arctic species remain particularly in areas such as the Baltic.

The seawater environment determines the species that live within it. The distribution of plants and animals depends on a number of factors, particularly *temperature*. Many marine animals are cold-blooded (poikilothermic); in other words the rate of all their life processes such as feeding, respiration and growth is increased by a rise in temperature and slowed down by a decrease. Most marine animals are acclimatised to the range of temperatures which they normally experience and will not survive if exposed to abnormal temperature conditions. The warm-blooded animals, notably the birds and whales in the marine environment, are able to control their body temperature independently of the environment and so they are less rigidly affected by sea or air temperatures. The effect of temperature on marine plants is variable; some species appear to have narrow tolerances whilst others can survive and grow in a wide range of temperatures: this is reflected by their distribution.

The seas of north-west Europe receive the warming influence of the North Atlantic Drift, an extension of the Gulf Stream. Unnaturally high temperatures, perhaps artificially induced by waste heat from industrial processes, can kill, or encourage other animals that will alter the balance of the indigenous population.

Another very important parameter of the marine environment is *salinity*. In most marine animals, apart from the vertebrates, the body fluids remain at a salt concentration equal to that of the surrounding water, although the exact composition of the salt content may be slightly different. The body fluids of marine animals are in contact with the surrounding seawater via the semi-permeable membranes which enable the animal to take up oxygen and eliminate carbon dioxide and waste products. When the external salt concentration changes, an imbalance is caused and different animals respond to this in different ways. In many instances, the body fluids adjust to match the external concentration, while in others, physiological mechanisms exist which enable the animal to maintain a constant level of salts in the body fluids despite external changes. In the open sea salinity remains remarkably constant but on the shore, in estuaries or in some enclosed seas, it may fluctuate or be considerably reduced. Under these conditions only those animals able to tolerate changes in salt levels within the body or those able to regulate their salt concentration will be able to survive. The fauna and flora of estuaries are thus reduced to those species which have evolved adaptations to continuously fluctuating salinities.

The salinity of oceanic water in the temperate zones remains at around 35 parts per thousand (ppt). Coastal waters are diluted by freshwater drainage from the land and salinities drop quickly in estuaries and enclosed seas receiving a freshwater inflow. Oceanic water bathes much of the coast of north-west Europe, but slightly reduced salinities are encountered in the English Channel and North Sea. The Baltic is brackish, the surface ranging from 15–20 ppt in the western regions to less than 5 ppt at the eastern end. In estuaries, marked changes in salinity occur each day with the tide, and apart from the reduction in salinity which is encountered on passing up an estuary there is often a marked layering of the water, so that fresh water flows out over the top of the more dense saline layers below.

The productivity of our seas is dependent on the quantity of *nutrient salts* available for the growth of plants, which support the production of herbivorous animals which, in turn, provide food for the carnivores at the top of the food web. The fertility of seawater changes with the seasons and considerable geographical differences exist. The major plant nutrients, nitrate and phosphate, find their way into seawater from the land, dissolved in the freshwater runoff, and in shallow waters by circulation from the seabed. The shallow coastal waters of the continental shelf are, therefore, considerably more productive than the offshore oceanic waters. Estuaries contain some of the most productive waters on earth. A seasonal cycle of production occurs in the sea. In the temperate zone during the winter months, dissolved nutrient levels are high due to the mixing of the waters by storms and increased freshwater runoff from the land, caused by high rainfall. With the coming of spring and the increase in the amount of sunshine, the minute plants start to grow and divide. At this time the water becomes very opaque and takes on a dark green or even brown appearance. The spring bloom lasts as long as the nutrients are available, but with the coming of the calm weather of summer these are diminished, plant growth stops and the dead cells, in which the nutrients are locked up, sink to the bottom. Usually in late summer a further bloom occurs when the first storms of the autumn turn over the water, bringing the nutrients once more to the surface. This second bloom declines with the onset of winter and the corresponding reduction in light level. This is the basic pattern but considerable variation occurs from year to year and place to place.

The *light absorption* by seawater is also a critical factor affecting the distribution of species. Sunlight falling on the surface of the sea is reflected to a greater or lesser extent according to height of the sun. Below, the surface light is rapidly absorbed, reaching 1 per cent of its surface value at 100 metres in clear oceanic water. In coastal waters, the 1 per cent level may be reached at 10–30 metres and in very turbid inshore waters and estuaries at 3 metres. Inshore waters tend to be clouded by plankton and suspended organic and inorganic matter. The amount of this will change with the season, sunlight, degree of freshwater runoff from the land, turbulence and with spring and neap tides.

Plants, which both produce organic matter by photosynthesis in the presence of light and utilise it for respiration, have a compensation depth at which production balances utilisation. Below this depth, light levels are insufficient to maintain photosynthetic activity at a sufficient level to prevent utilisation outstripping production, and so growth cannot occur. Seaweeds will not be found below this level and phytoplankton carried down or sinking below it will die unless carried back towards the surface. The importance to seaweeds of differential absorption of the various wavelengths of light is mentioned in Chapter One.

Oxygen is only moderately soluble in seawater and the solubility decreases at higher temperatures and also at higher salinities. The amount of oxygen present depends upon the balance between the various sources of input and uptake by animals and plants for respiration and the requirements of bacterial decomposition. The primary source of oxygen in the sea is the atmosphere and dissolution is aided considerably by agitation of the sea surface by wind and wave action. The second source of oxygen is the plants, both seaweeds and phytoplankton, which produce it during the process of photosynthesis. In the depths of the sea oxygen levels are kept up by circulation of oxygenated water from the surface and so animal life is found in even the deepest ocean trenches. In certain isolated basins which are more or less cut off from the general circulation of the oceans, oxygen depleted layers may be present near the bottom. Such is the case in the 'polls' which occur at the head of Norwegian fjords and are only connected with the rest of the water body via shallow sills. Dead zooplankton and phytoplankton sinking from the surface water in these polls are decomposed by bacteria which need oxygen for the process. As there is only limited circulation from the surface, an anoxic layer develops. A small tidal range and freshening of the surface waters, causing a layer effect, accentuates the situation and if the deoxygenated bottom water is brought to the surface, catastrophic mortality of animals in the upper layers may ensue.

Carbon dioxide (CO_2) is of great biological significance, as it is essential for photosynthesis by plants. It is present in seawater at about sixty times the concentration as in the atmosphere because not only is it present as the dissolved gas but it also forms carbonic acid (H_2CO_3), bicarbonate (HCO_3) and carbonate (CO^2_3). CO_2 in solution is used by plants for photosynthesis, causing a shift in the equilibrium between these compounds, and bicarbonate is drawn upon to supply the CO_2 which has been removed.

Acidity or *pH* of seawater is a measure of the concentration of hydrogen ions. pH is measured on a scale of 1–14, where a pH of 7 indicates neutrality and higher values indicate increasing alkalinity and lower values indicate increasing acidity. Seawater at the surface is usually slightly alkaline with a pH of between 8·1 and 8·3. The carbon dioxide system plays an important part in regulating pH of seawater and makes it a fairly good buffer so that pH changes little, even on the addition of small quantities of strong alkali or acid. Further addition will result in the exhaustion of the buffering capacity, but under normal conditions the pH of seawater remains remarkably constant.

The addition of carbon dioxide to seawater increases the acidity, that is it depresses the pH, and removal has the opposite effect. Where respiratory activity is high, such as in waters where there are high levels of decomposing organic matter, pH levels will be low. Conversely, where photosynthetic activity is high, for instance, in high light conditions and where nutrients are abundant, then pH will be increased. The polluting effects of untreated sewage or eutrofication may thus influence pH to a certain extent. Where pH levels exceed the normal limits, the health of animal life may suffer.

There is a section in Chapter Five which deals with ways of measuring temperature, salinity, light absorption, dissolved oxygen and pH value, so that the yachtsman/naturalist may assess a particular environment.

THE CLASSIFICATION OF PLANTS AND ANIMALS

The plants and animals described in this book have all been given their scientific names and a common name where one exists. The use of long latinised names may seem tiresome at first to the amateur naturalist, but common names are of little real value in identification: they are often inconsistent and are rarely universally accepted. (Indeed, the observant marine naturalist may discover species which are so unusual that they have never been given common names at all.) All living organisms are named according to the binomial system of nomenclature introduced in the eighteenth century by Linnaeus, the Swedish naturalist. Under this system, each organism receives two names: the first is the generic name, written with an initial capital letter, and the second is the specific, or trivial, name, which is not capitalised. An example is *Ostrea edulis*, the scientific name for the common European oyster. The trivial name is usually followed by the name of the first person to describe the species; this name may be placed in parentheses if the species has subsequently been placed in a new genus—for example, *Pecten maximus* (Linnaeus), commonly called the great scallop.

The animal and plant kingdoms are divided into six major categories of descending magnitude:

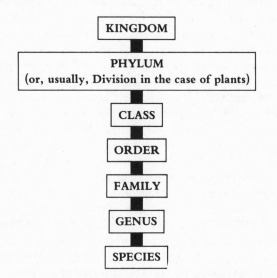

These categories are often further subdivided into, for instance, subphyla, subclass, and so on. The plants and animals described at the end of Chapters One, Two and Three have been placed in their respective phylum or division. Further subdivision has been given only where necessary, to avoid the descriptive sections becoming too cumbersome.

It is important that the naturalist should familiarise himself with the distinguishing features of the major phyla of marine animals and plants, as this will enable him to start off on the right track when identifying new specimens. Unless all the features of an organism are considered carefully, it is possible to confuse even the phyla to which an organism belongs: for many years it was believed that the barnacles belonged to the phylum of molluscs but it is now known that they are crustaceans—a fact which is confirmed beyond doubt when the larval stages of these animals are considered. A short description of each phyla covered by this book is given here.

Plant Kingdom

Algae
By far the most successful division of plants on the shore and in the sea are the

algae. The group includes the red, brown and green seaweeds or macrophytes of the shore and shallow water, and the many different kinds of single-celled planktonic forms. The seaweeds are very simple plants; their bodies are not differentiated into true roots, stems or leaves and they lack a vascular system (i.e. channels for conducting liquids). Typical seaweeds are composed of a holdfast used to attach the weed to the seabed, a frond and sometimes a stalk or stipe. Reproduction in the group is complicated but essentially two methods are employed: asexual reproduction by spores and sexual reproduction by the production of gametes. Some species employ both methods and they alternate between generations so a sporophyte generation gives rise to a gametophyte which gives rise to a sporophyte, and so on. All seaweeds possess the green photosynthetic pigment chlorophyll which enables the plant to make use of the energy of sunlight, but this is masked in the brown seaweeds by the brown pigment fucoxanthin and in the red seaweeds by the red pigment phycoerythrin.

The single-celled algae dominate the planktonic plants or phytoplankton. The most prominent members are the diatoms with their delicately sculptured siliceous skeletons which enclose the protoplasm of the cell like a pill-box. They have evolved into a fantastic diversity of shapes and sizes and often individual cells may link together to form chains or other shapes. They usually reproduce by simply dividing into two but because of the rigid pill-box-like construction of the skeleton, each division results in a size reduction of the cell. After a number of divisions, the siliceous valves are thrown off, sexual reproduction may occur, and a soft-bodied auxospore is produced which increases the size of the cell before a new skeleton is formed once more. The diatoms also produce resting spores which sink to the bottom when conditions are unsuitable for growth.

No less common than the diatoms, but often less conspicuous because of their smaller size, are the flagellates. There are many different kinds but all are characterised by the possession of one or sometimes more whip-like flagellae which these cells use to propel themselves through the water.

Lichens

Moving out of the sea and on to the shore we find the lichens which are dual plants composed of fungus and algae combined in a single body. Reproduction is usually achieved by fragments of the lichen body, containing a fungal and an algal element, breaking off and being carried away to a suitable environment by the wind.

Bryophyta (Mosses)

Close to the top of the shore various mosses are often to be found. These belong to the division known as the Bryophyta. The tissue is more differentiated than in the algae, but there is no true transport or vascular system within the plant. As with some of the algae, reproduction is both sexual and asexual and generations exhibiting one or other mode alternate. It is the sexual or gametophyte generation which is usually conspicuous.

Pteridophyta

After the mosses come the pteridophytes. These are a major division of vascular plants containing the ferns and their allies, the horsetails, clubmosses and quillworts. They are largely terrestrial and have true stems, leaves and roots. Like the mosses there is alternation between sexual (gametophyte) and asexual (sporophyte) generations and in the case of the pteridophytes it is the sporophyte generation which is dominant and conspicuous. *2599994*

Gymnospermae (Conifers)

The coniferous trees belong to this primitive group of woody seed plants which bear their seeds exposed, usually in a cone, rather than in a fruit.

Angiospermae (Flowering Plants)

The final division of plants usually associated with the sea are the angiosperms or flowering plants. These include the familiar grasses, herbs, shrubs and trees. They are mostly confined to the shore and above but a few species are truly marine. The angiosperms display a high degree of tissue differentiation with roots, stem and leaves, and they possess a vascular system and seed-bearing flowers.

Animal Kingdom

Porifera (Sponges)

The simplest animals dealt with in this book are the sponges which belong to the phylum Porifera. The sponge body consists of a single cavity with many small inhalent pores, through which water is drawn, and a single exhalent opening known as the osculum. A number of units based on this simple body plan may unite to form a complex sponge. Water currents required for suspension feeding and respiration are created by the action of the choanocytes which are flagellated cells lining the body cavity. There are no specialised tissues in the sponge body but eggs and sperm are produced which unite during fertilisation to form free-swimming larvae which, if they settle on a suitable substrate, will develop into new sponges.

Coelenterata

The coelenterates, or Cnidaria, include the sea-firs, sea-anemones, corals and jellyfish. They have simple sac-shaped bodies enclosed by a wall of two layers of cells separated by a jelly-like mass. This may be extensive in forms such as the jellyfish. There is a single opening to the body, usually termed the mouth, and this is surrounded by a ring of tentacles which bear stinging cells. Individuals are termed polyps or, in the case of jellyfish, medusae, and there is often a medusoid stage in the life cycle.

Ctenophora (Sea-gooseberries)

The phylum Ctenophora contains the sea-gooseberries or comb-jellies. These animals resemble the coelenterate medusae in habit and appearance, and in the possession of a jelly-like mass which separates the two layers of the body wall. They differ from the coelenterates, however, in possessing true muscle cells and rows of fused cilia on the outer body surface. There are usually two long tentacles which do not sting but capture the prey by a lassoing action.

Platyhelminthes (Flatworms)
The flatworms are not particularly well represented in the sea. They are simple animals, often leaf-shaped, with simple eye spots and tentacles at the anterior end. They usually glide along the bottom by the action of the many cilia on their underside.

Nemertina (Ribbon worms)
Ribbon worms, as the name suggests, are ribbon-shaped and often extremely long. The body is unsegmented and consists of three layers of cells surrounding the gut. The mouth possesses a proboscis which is used to capture prey and there are simple eye spots at the head end of the worm.

Annelida (Segmented worms)
This phylum represents a major advance in the evolution of the animal body plan. The body consists of three layers of cells and the middle layer is divided into two by a fluid-filled body cavity or coelom. The body is divided into a number of segments and there is a well-defined head bearing sense organs and a simple brain. There are three classes within the phylum Annelida and the first of these, the polychaetes or bristle worms, are the most abundant in the sea. The earthworms and their allies, or oligochaeta, and the leeches, or hirudinea, are more usually associated with terrestrial and freshwater habitats, although there are a few marine representatives.

Priapuloidea
This is a minor phylum of gherkin-shaped worms with a tassel-like tail and an eversible proboscis. The body is segmented externally and it possesses a true coelom.

Chaetognatha (Arrow worms)
The arrow worms are a purely planktonic phylum. They are all small and have a torpedo-shaped, bilaterally symmetrical body with side fins and a tail fin. There is a mouth at the anterior end with a series of strong spines used for grasping prey.

Mollusca (Molluscs)
This is a well-known phylum containing the chitons, snails, slugs, bivalves, tusk shells, octopus, cuttlefish and squid. The mollusc body is bilaterally symmetrical and unsegmented; it has three layers of cells and possesses a coelom. The body is

covered by a mantle which secretes the shell in those species where one is present, and it encloses a cavity in which the gills lie. In many molluscs there is a muscular foot which performs a variety of functions in the different groups. Indeed, it is clear from the above list that the basic molluscan body has successfully evolved into a variety of very different forms.

Arthropoda

This is the largest phylum in the animal kingdom. These animals are bilaterally symmetrical, segmented animals with bodies consisting of three cell layers; there is a coelom and often a large blood space. Some or all segments bear paired appendages including mouthparts and there is a hard, jointed exoskeleton composed of chitin. There is also a well-developed nervous system. The most important class of arthropods in the sea is the Crustacea which contains such forms as the barnacles, shrimps, prawns, crabs and lobsters. Another less important class is the Pycnogonida, or sea-spiders, which despite their name are not closely related to the true spiders. The insects and true spiders, although so successful on land, are virtually absent from the marine environment.

Polyzoa

The Polyzoa, otherwise known as the Bryozoa or Ectoprocta, is a largely inconspicuous phylum consisting of small, sessile, colonial animals, referred to as zooids, which live in a secreted, box-like case or zooecium. This is sometimes hardened by chalky deposits laid down by the body wall. There is a true coelom and the mouth is surrounded by a ring of tentacles known as the lophophore. The anus always lies outside the lophophore; this feature distinguishes the polyzoa from the sea-firs which they resemble superficially.

Echinodermata

This is a very distinct phylum containing the feather stars, starfish, brittle stars, sea-urchins and sea-cucumbers. The most obvious feature of the echinoderms is their pentamerous, or five-sided, symmetry. The body is composed of three cell layers; the ectoderm, or outer cell layer, covers the calcareous skeleton which consists of a series of plates. These are rigidly connected in the case of the sea-urchins. The spines of sea-urchins are further skeletal elements which project through the ectoderm. The members of this phylum possess a unique water-vascular system. This is a type of hydraulic system used to operate the tube feet which are provided with suckers and are used for locomotion in the starfishes, sea-urchins and sea-cucumbers, and for filter-feeding by the feather stars.

Hemichordata

This small phylum is exclusively marine. The animals possess bilaterally symmetrical, worm-like bodies which are divided into three distinct zones. Tentacles may be present on the middle zone. Both solitary and colonial forms occur.

Chordata

This phylum contains animals with a single, hollow, dorsal nerve chord, a true coelom, gill slits and a strong flexible rod, or notochord, which stretches the whole length of the body. There is also a post-anal tail. There are two subphyla within the Chordata; the first of these is the Vertebrata which includes the cyclostomes, lacking jaws, the fishes, birds, mammals, reptiles and amphibians. The subphylum Urochordata contains three invertebrate classes of chordate. The ascidians, tunicates or sea-squirts, possess the hollow dorsal nerve chord only in the tadpole-like larva and in the adult form are usually sessile and covered with a cellulose tunic. The Larvacea or Appendicularia are pelagic tunicates which retain the larval tunicate features of a dorsal notochord in the trunk and tail. They are surrounded by a transparent gelatinous house through which water for suspension feeding and respiration is drawn by the movements of the tail. The third class, Thaliacea, resemble floating, jelly-like barrels. A hollow dorsal nerve chord and tail are only present in the larva of this group.

Part One

CHAPTER ONE

The Seashore

The term 'seashore' is difficult to define precisely, but for our purposes we will consider it here to encompass the area extending from low tide level to the upper limits of the tide and beyond, to include rocky cliffs and sand dunes. The area is divided into a number of zones. The intertidal or littoral zone is that region bounded on the landward side by the level of extreme high water spring tide (EHWS) and on the seaward side by extreme low water springs (ELWS). The splash zone lies above EHWS. Other levels recognised are mean high water of spring tides (MHWS), mean high water of neap tides (MHWN), lowest high water neap tides (LHWN), mid tide level (MTL), highest low water of neap tides (HLWN), mean low water of neap tides (MLWN) and mean low water of spring tides (MLWS). For biological purposes the shore may be broadly divided into the upper shore, which is above the average high tide level (AHTL), the middle shore from between this and the average low tide level (ALTL), and the lower shore between this and ELWS. AHTL and ALTL are shown by dotted lines in Fig. 1.

TIDES

The average tidal range in north-west Europe is 5 metres at springs and 3·8 metres at neaps, but the configuration of the land has an important effect on the tides of all the countries bordering the English Channel and the North Sea. The range is increased in funnel-shaped inlets such as the Bristol Channel: at Chepstow the range is 14 metres at springs and 7 metres at neaps. On the other hand, conflicting water movements may reduce the tidal range, so that between Portland and Swanage on the English coast it is a mere 1·7 metres while across the Channel at St Malo it is some 12 metres. In the Baltic, which is almost completely enclosed by land, and on the coasts of southern Scandinavia, the tides are scarcely noticeable.

Two tidal waves reach the coasts of north-west Europe every lunar day (24 hours and 50 minutes). They approach from the west and are divided by the land mass of Ireland. The southern half of the wave divides again at the south-west tip of England, one part passing up the Bristol Channel and the other into the English Channel. The northern half sends a minor wave down into the Irish Sea and travels round the north of Scotland and down into the North Sea. Taking 12 hours to reach the southern end of the North Sea it meets the succeeding wave, passing up the English Channel, at the Straits of Dover. The times of the tides thus vary widely around the coasts.

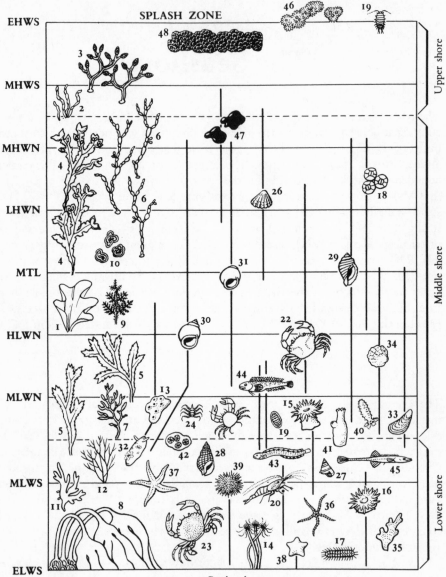

SPLASH ZONE

EHWS

MHWS

MHWN

LHWN

MTL

HLWN

MLWN

MLWS

ELWS

Upper shore

Middle shore

Lower shore

Rocky shore

Plants and animals not drawn to scale

EHWS Extreme high water spring tides   **HLWN** Highest low water neap tides
MHWS Mean high water spring tides   **MLWN** Mean low water neap tides
MHWN Mean high water neap tides   **MLWS** Mean low water spring tides
LHWN Lowest high water neap tides   **ELWS** Extreme low water spring tides
MTL Mid tide level

Key to Species
1. *Ulva lactuca*: Sea-lettuce
2. *Enteromorpha inteslinalis*: Grass kelp
3. *Pelvitia canaliculata*: Channelled wrack
4. *Fucus vesiculosus*: Bladder wrack
5. *Fucus serratus*: Serrated wrack or toothed wrack
6. *Ascophyllum nodosum*: Knotted wrack or egg wrack
7. *Halidrys siliquosa*: Pod weed
8. *Laminaria digitata*: Oarweed
9. *Corallina officinalis*: Coral weed
10. *Lithothamnion incrustans*
11. *Rhodymenia palmata*: Dulse
12. *Griffithsia flosculosa*
13. *Halichondria panicea*: Crumb-of-bread sponge
14. *Tubularia indivisa*: Sea-fir
15. *Actinia equina*: Beadlet anemone
16. *Tealia felina*: Dahlia anemone
17. *Lepidonotus squamatus*: Scale worm
18. *Balanus balanoides*: Acorn barnacle
19. *Ligia oceanica*: Sea-slater
20. *Halaemon serratus*: Common prawn
21. *Porcellana platycheles*: Hairy porcelain crab
22. *Carcinus maenas*: Common shore crab
23. *Macropipus puber*: Fiddler crab or velvet swimming crab
24. *Nymphon gracile*: Sea-spider
25. *Lepidochitona cinereus*: Grey chiton shell or coat-of-mail shell
26. *Patella vulgata*: Common limpet
27. *Calliostoma zizyphinum*: Painted topshell
28. *Nassarius reticulatus*: Netted dogwhelk
29. *Nucella lapillus*: Common dogwhelk
30. *Littorina littorea*: Edible periwinkle or common periwinkle
31. *Littorina littoralis*: Flat periwinkle
32. *Archidoris pseudoargus*: Sea-lemon
33. *Mytilus edulis*: Common mussel
34. *Anomia ephippium*: Common saddle oyster
35. *Flustra foliacea*: Hornwrack
36. *Ophiothrix fragilis*: Brittle star
37. *Asterias rubens*: Common starfish
38. *Asterina gibbosa*: Cushion-star or starlet
39. *Psammechinus miliaris*: Green sea-urchin
40. *Cucumaria normani*: Sea-cucumber
41. *Ciona intestinalis*: Sea-squirt
42. *Botryllus schlosseri*: Star sea-squirt
43. *Pholis gunnellus*: Gunnel fish or butter fish
44. *Blennius pholis*: Common blenny or shanny
45. *Spinachia spinachia*: 15-spined stickleback
46. *Xanthoria parietina*
47. *Verrucaria mucosa*
48. *Verrucaria maura*

Fig. 1. Zonation of plants and animals on the rocky shore. Species listed according to their classification.

Spring tides occur twice each month, two days after each full and new moon, when the gravitational pull of the moon and the sun is in a straight line. The neaps come in between, after the first and third quarters of the moon, when the direction of pull of the moon and sun is at right angles. The tides with the biggest range are twice yearly at the March and September equinoxes. Climatic factors, particularly winds and conditions of high or low barometric pressure, may have significant effects on the tidal range; thus an onshore wind associated with low pressure may lead to exceptionally high tides and an offshore wind with high pressure will have the opposite effect.

Tidal movement creates currents in coastal areas and these may exert a considerable influence in supplementing or inhibiting the transporting effects of other sea currents. The turbulence set up by tidal currents and their changes in direction may affect life on the shore, as it prevents the settlement of suspended particles from the water and so causes increased turbidity or cloudiness.

The seashore provides a particularly difficult environment for plants and animals to colonise because of the very variable conditions which exist there. When the tide is high the shore is bathed in water of reasonably constant temperature and salinity, but when it recedes all life is exposed to the effects of desiccation and the blazing hot sun in the summer or the keen frosts of winter. Those animals and plants living in pools may find the salinity raised by evaporation or drastically lowered by heavy rainfall. In stormy weather, waves, often carrying stones, dash against the rocks or churn up the sand. Apart from very unstable pebble or shingle areas, however, our shores are inhabited by a fascinating variety of organisms which often provide the naturalist with his first taste of marine biology, because of the ease with which this habitat may be studied when exposed by the retreating tide.

ZONATION

The seashore has been colonised almost entirely by specialised members of the marine bottom fauna and flora which have become adapted in varying degrees to the effects of exposure to the air. Very few animals, apart from a few inconspicuous insects, have invaded the shore from the land. Adaptation to life on the shore may involve simple tolerance to the adverse conditions which exist while exposed by the ebb tide, or special physiological mechanisms may be present such as the ability of certain bivalve molluscs to carry out their essential life processes in the absence of oxygen, enabling them to remain closed at low water. Some animals have behavioural tricks such as burrowing or hiding under rocks at low water and so they avoid the worst of the effects of being uncovered by the tide. There is also a host of animals such as fish and crabs which migrate into the intertidal zone at high tide and move offshore again only during the ebb.

Because there is approximately a six-hour difference between high and low water, the amount of time which an animal or plant is exposed to the air depends on how far up the shore it lives. As a result of the varying degrees of exposure which exist, we find a zonation pattern whereby organisms best adapted to long periods of exposure occur high on the shore and those less well adapted live lower down. Zonation is most apparent on rocky shores, because the animals and plants are mostly visible on the surface of the rocks. On shallow sloping shores, the different zones with their characteristic species may extend over several metres, whereas on a steeply sloping shore, the zones are compressed.

On sandy and muddy shores animal life occurs in distinct zones, much as with rocky shores, but these zones are less obvious because nearly all the species live in burrows beneath the sand, at least at low tide. Apart from the behavioural responses of these animals, adaptation to aerial exposure is generally far less developed as marine muds and sand tend to hold a large quantity of water and conditions may be very damp just below the surface even when the tide is out. In fact, zonation on sandy and muddy shores is probably determined largely by the amount of time that different animals need to be covered in order to obtain adequate food. On sandy and muddy shores there is usually a marked paucity of large seaweeds although there may be a few in very sheltered areas or where there are occasional large stones or boulders for attachment.

WAVE ACTION

The amount of wave action which our shores receive varies considerably around the coasts of north-west Europe, which range from the steep rocks of north-west France, western Ireland and Scotland, receiving the full force of the Atlantic swell, to the fjords of Norway, the sheltered shores of the Frisian Islands and the quiet estuaries of the Dutch coast and eastern England. On rocky shores wave action is a major factor governing the type of population which is present. On sheltered shores brown seaweed growth is luxuriant and there is usually only a narrow gap between the marine algal population and the terrestrial vegetation above. In wave-swept shores the algal growth is stunted; it is separated from the terrestrial vegetation by a zone of bare rock and the shore is often characterised by communities dominated by mussels and small red algae. In many intermediate situations, barnacles often dominate the community. Sandy and shingle shores exposed to wave action are generally rather unstable and support little life; those in sheltered areas however may be very rich, although to find the animals on those shores involves strenuous digging. Muddy shores are usually found only in quiet, sheltered areas such as in estuaries or sea lochs.

THE ROCKY SHORE

Some of the most spectacular coastal scenery for the yachtsman cruising close to shore lies on the westerly extremes of north-west Europe: the coasts of Norway, western Scotland and the Scottish islands, the west coast of Ireland, England, the Brittany coast and that of northern Spain. Here the great rollers of the ocean meet rugged cliffs and shores of ancient hard rocks which are resistant to erosion and the forces of the sea. These shores create a challenge to the yachtsman as faulty navigation and seamanship can have serious consequences where the bottom is scattered with razor sharp rocky outcrops, but from a naturalist's point of view the ecology of these areas has many attractions. Sea cliffs provide a habitat for a wide variety of plants, and during the spring seabirds come to these areas in large numbers to nest. The intertidal zone on rocky shores provides a habitat for luxuriant growth of seaweeds and animal life, especially in rock pools. The clear water usually encountered near rocky shores provides ideal conditions for snorkelling and Scuba diving.

On the rocky slopes and cliffs above the level of the highest tides there is little opportunity for plant growth, but here and there in the rocky depressions a thin layer of topsoil collects and provides a place for vegetation to take root. There are a number of flowering plants which tolerate the salt spray and harsh conditions of these sites and they are characteristic of rocky coastal areas. The best known of these flowers is probably the sea-pink or thrift (*Armeria maritima*) (**II, 1**) which is commonly found growing in tufts on shallow soil or even in cracks in the rocks at the top of the shore and on coastal cliffs. It is common from France to as far north as Iceland. The rock sea-lavender (*Limonium binervosum*) (**I, 6**) comes from the same family as the sea-pink; like its close relative it favours rocky areas and sea cliffs behind the shore. Rocky cliffs and patches of shingle at the top of the shore provide a habitat for the wild cabbage (*Brassica oleracea*) (**I, 9**). This large plant has broad leaves which are covered with a bluish-green bloom. Scurvy grass (*Cochlearia officinalis*) (**I, 1**) occurs in similar situations. This plant, with its conspicuous white flowers, used to be collected by mariners in the days of sailing ships as a protection against scurvy—hence the name. The leaves are a valuable source of ascorbic acid (vitamin C). The samphire (*Crithmum maritimum*) (**I, 8**), or rock samphire as it is sometimes known, is named after St Peter, the patron saint of fishermen. It used to be known as crestmarine. It is common, rooted in crevices in rocks, and the fleshy leaves are sometimes gathered and pickled in brine. On cliff ledges and also on the top of shingle beaches the sea-campion (*Silene maritima*) (**I, 7**) grows over large areas in cushion-like masses. On shingle shores it has a stabilising effect. The woody rootstocks survive for several years and produce annual shoots which die back at the end of each season.

The Baltic shores and islands are particularly interesting from the ecologist's point of view, and to treat them fully would require a separate book. There are, however, certain plants which seem to typify the area, which is unique in western Europe. The larger islands which support vegetation have been colonised mainly by three species of tree: the Norwegian spruce (*Picea abies*), the pine or Scots pine (*Pinus sylvestris*) and the juniper (*Juniperus communis*) (p. 63). They are usually rather stunted in this area. The juniper grows either as a small tree or as a bush, and on those islands where the spruce and pine are absent, the low, stunted bush form predominates. These conifers are able to establish a foothold on rocky islands in humus (which has formed in a shallow basin) that may be as little as 8 cm deep.

Cruising around the Baltic islands in the summertime the sailor could expect to see a number of other plants in addition to the trees. The lady fern (*Athyrium filix-femina*) (p. 61) prefers shady rocks and damp areas, and cotton grass (*Eriophorum polystachion*) (p. 65) also likes wet boggy places. Chives (*Allium schoenoprasum*) (p. 65) is very local in its distribution but is often to be found growing in tufts, especially where the ground is rocky. It is sometimes cultivated for flavouring and is excellent in salads and omelettes. The wild or tricolour pansy (*Viola tricolor*) (p. 61) varies in colour from violet to mauve and yellow on the same flower, to three shades of mauve or mauve and white in the same clump. Another flower which is very showy on the Baltic islands is the scentless chamomile (*Tripleurospermum maritimum*). It is to be seen in flower from July to September.

There are many features of the ecology of rocky shores which are common to all areas of north-west Europe and the description given below of the intertidal zone is intended to be as general as possible. It should be remembered, however, that many areas have their own special features—it is this which makes the subject of shore ecology so fascinating, and there is ample opportunity for the mobile yachtsman to make comparative studies of the shores of different areas.

At the very top of the shore above the seaweeds, any rock surface will usually be covered with lichens. *Xanthoria parietina* (var. *ectanea* : **I, 3**) is orange in colour; it has a rough leafy texture and may form a distinct belt on the extreme upper shore. It is often associated with other orange lichens of the genus *Caloplaca*. Below this, and sometimes overlapping with *Xanthoria* on shores where the splash zone is wide, the black, lifeless-looking lichen *Verrucaria maura* (**I, 4**) coats the rocks. This species is often so extensive it may resemble the dried remnants of an oil spill and it occurs nearly everywhere, being scarce only where the rock is soft, crumbling or very cracked or where loose shingle or mobile boulders are present. Further down on the middle shore, separated from the other lichens, the greenish-black lichen *Verrucaria mucosa* (**I, 5**) may be found. The glossy surface of this species gives it a quite different appearance from *V. maura* and its lower position on the shore means that it should not be confused. It occurs among the upper level of barnacles on exposed shores but is likely to be more extensive on sheltered vertical faces or on boulder shores where barnacles and seaweeds are scarce.

Below the lichen zone we can expect to see some considerable differences in the dominant plants and animals present, depending on the degree of exposure of the shore to wave action. In sea lochs, fjords or enclosed bays which are fairly well sheltered from the waves the most obvious sign of life will be the brown seaweeds. These weeds, which make walking over the rocks so treacherous, provide a wonderful damp habitat for many animals and are themselves a fascinating source of study for the naturalist.

It is the seaweeds which display zonation most strikingly in the intertidal area. The first seaweed usually growing at the top end of the shore at the upper limit of the tidal range is the brown fucoid known as channelled wrack (*Pelvitia canaliculata*) (**II, 6**) which forms a dense tufted growth over the rocks. It occurs in most areas, except the Baltic or where there is considerable exposure to wave action. In its position on the shore, this weed spends much of its time exposed to the air and on hot days dries out almost completely and goes very dark brown or almost black in colour. Further down, the shore is dominated by a variety of other fucoid algae, the best-known example of which is the bladder wrack (*Fucus vesiculosus*) (**II, 5**). The fucoid zone can differ from site to site depending again on the degree of exposure and also on the type of profile—whether the shore is steeply sloping or falls away in a broad sweep. The uppermost wrack is often the spiral wrack (*Fucus spiralis*) (p. 61). This species lacks bladders but has conspicuous rounded reproductive bodies on the end of the branches which are usually twisted for at least some of their length. Below this, the bladder wrack may occupy a broad belt, in many situations covering much of the middle shore, although where there is considerable shelter from wave action, another fucoid, the knotted or egg wrack (*Ascophyllum nodosum*) (**II, 3**), with its large bulbous bladders, may predominate. Below these weeds we can usually expect to find a narrow strip of the toothed wrack (*Fucus serratus*) (**II, 2**). This species, which lacks bladders and has a fine serrated edge to its fronds, is often heavily covered by the coiled white calcareous tubes of the small worm *Spirorbis borealis* (**II, 2**).

At the very lowest limits of the intertidal zone, and often only uncovered by spring tides, is the *Laminaria* or strap-weed zone. There are three species of *Laminaria*: *L. digitata* (oarweed) (**II, 4**), *L. hyperborea* and *L. saccharina*. These are the largest weeds found in European seas, and as we shall see in the next chapter (where *L. saccharina* is illustrated) they extend for some distance down into the sublittoral region. The stipes of these great weeds often provide a site for the attachment of the red seaweed, dulse (*Rhodymenia palmata*) (**II, 7**), and the holdfasts provide a sheltered habitat for a great many small worms, crustaceans and molluscs. On exposed shores the dabberlocks (*Alaria esculenta*) (p. 61) is often present among the *Laminaria* although it is easily distinguished by a very obvious midrib which extends the full length of the frond. This species tends to have a northern distribution and becomes scarce at the latitude of the English Channel and further south.

There are many more seaweeds which occur among the dominating brown weeds of the different zones. On the middle and lower shore in quiet channels, especially where there is shingle and sand, the pod weed (*Halidrys siliquosa*) often grows and it may replace *Laminaria* in this kind of situation. It is immediately recognisable by growths resembling the seedpods of some land plants, which grow on the tips of the side branches. In certain situations other algae may become predominant and occupy distinct zones of their own. Where there is any fresh water running onto the shore in the vicinity of a stream, or in estuaries, the green weed *Enteromorpha intestinalis* (**II, 15**), with its small inflated fronds, forms a distinct covering to the rock. Another very common green weed is the sea-lettuce (*Ulva lactuca*) (**II, 29**), which usually inhabits pools and damp areas on the upper and middle parts of the shore.

The red seaweeds are an enormous group of beautiful plants, many of which are difficult to identify for certain. Attached to rocks in pools, *Griffithsia flosculosa* is often common and so too is the pink calcareous coral weed (*Corallina officinalis*) (**II, 11**). This latter plant is widely distributed and is very tolerant of severe wave action. It is especially abundant on headlands and in surge gullies on exposed Atlantic coasts and will grow in dense carpets well above the *Laminaria* zone in situations where little else is present. In some rock pools, particularly in exposed places, the encrusting red algae *Lythophyllum* may coat the rocks to give the appearance of a layer of pink icing sugar (*L. incrustans*: **II, 32**). The edge of the growth of this algae is often sharply marked at the level in the pool to which the water level drops on hot days.

Many of the animals of the rocky shore are inconspicuous and well hidden at low tide and so they have to be searched for carefully. There are, however, a number of species which are very noticeable and may themselves form distinct zones on the rocks. Some animals are characteristic of certain types of shore; for example, barnacles such as the acorn barnacle (*Balanus balanoides*) (**II, 22**) and the star barnacle (*Chthalamus stellatus*) often occur in a very obvious band on the upper and middle areas of shores receiving moderate wave action, where the rock is reasonably hard. Barnacles are of course common on all shores, but in sheltered areas they often tend to be obscured by the presence of seaweeds. *Balanus balanoides* is a northern species distributed in the Atlantic, the English Channel, the North Sea and western parts of the Baltic, and where it is the only barnacle present it occupies a zone from the level of high water neap tides downwards. *Chthalamus stellatus* on the other hand is a more southerly species extending from the Mediterranean into the Atlantic and the English Channel. It is interesting that where the ranges of the two species overlap, in the southern parts of the range of *B. balanoides*, *C. stellatus* is always found above it; *B. balanoides* is then distributed from the middle shore downwards.

Where softer rocks such as chalk or sandstone are present, the common limpet (*Patella vulgata*) (**II, 19**) may dominate the upper and middle level of exposed shores, and although less dominant on harder rocks it is usually common. The limpet is a gastropod which has lost its coiled shell in the course of evolution and has been left with a simple cone-shaped shell. The elliptical edge of the shell always perfectly fits the patch of rock where the limpet makes its home, and the shape of the shell is well designed to withstand the pounding of the waves. At high tide it moves away from its home and grazes on the algal covering of the adjacent rock surface, using its rasping tooth-like radula. As the tide recedes the limpet then returns to its own patch of rock, where it must orientate itself correctly to fit against the surface and so avoid desiccation. The limpet is able to cling tightly to the rock by means of its foot; on soft rocks this action gradually wears away the surface and a 'limpet scar' is produced.

On exposed Atlantic shores with a gentle slope the common mussel (*Mytilus edulis*) (**II, 23**) may predominate, but where the slope of the shore is too steep it will tend to disappear and be replaced by barnacles. The mussel avoids desiccation at low tide by tightly closing its shell and trapping sufficient seawater to keep the tissues moist while uncovered. It is attached to the rock by means of byssus threads, secreted by a gland in the foot, which stick firmly to the surface and hold the mussel secure against the battering of even the most powerful waves. High up the shore mussels are usually tiny and may only attain a size of 10 mm or so, but lower down they are larger. Indeed the fisheries for mussels on the Dutch, Danish and West German coasts largely exploit populations which live at or below low water mark, on muddy gravel, and these are fished by dredging. The mussels' habit of byssal attachment makes them very amenable for mariculture. In Brittany they are cultured on *bouchots* or piles driven into the ground in areas where settlement of juvenile mussels is high, and in the rivers of northern Spain they are grown on ropes suspended from rafts, built especially for the purpose or from the hulks of old boats which have been adapted for the job.

There are several common snails which are obvious on most rocky shores. The predatory dogwhelk *Nucella lapillus* (**II, 16**) is found on all north-west European shores except in the Baltic. It feeds largely on barnacles and mussels which it attacks by drilling through the shell and extracting the soft parts within. The edible or common periwinkle (*Littorina littorea*) (**II, 21**) is ubiquitous on shores from mid-tide level downwards. It grazes small algal growth from the surface of the rock while slowly crawling over the surface. Although they are mobile, periwinkles maintain their position on the shore by following U-shaped feeding paths, using the sun to guide them. When starting to crawl the periwinkle may move either towards or away from the sun, and after a while the direction is reversed to bring it back close to its starting position. On vertical surfaces such as rock faces or groynes the periwinkle responds to gravity, moving downwards at the beginning of its migration and then reversing direction to return to the starting position. Periwinkles remain stationary for much of the time; when dried out at low tide, a small amount of mucus is secreted around the entrance of the shell, and as this dries it seals the animal in and prevents further desiccation. It is probably the action of wave movement with the rising and falling of the tide which stimulates activity.

The common periwinkle lives on a part of the shore where it spends much of its time submerged, and its mode of reproduction has evolved to suit this way of life. Tiny capsules containing one to six eggs are shed and these hatch into free-swimming planktonic larvae, which after some time in the plankton will settle on the shore while the tide is high. The rough periwinkle (*Littorina saxatalis*) (p. 73), which lives higher on the shore—just below high tide level—has evolved a different strategy. Because it spends so little time submerged, releasing eggs into the plankton would be far less satisfactory and so it retains its eggs and larvae in a brood chamber within the egg duct; these develop to a crawling stage before they are released. There is therefore no chance that they could be lost at sea. Strangely enough the small periwinkle (*Littorina neritoides*) (p. 73), which lives even further up the shore (at high water mark and above), produces floating eggs and free-swimming larvae. These are shed at fortnightly intervals throughout the winter at high water spring tide while the adults are submerged.

At the very top of the shore the longshoreman may see the sea-slater (*Ligia oceanica*) (**I, 11**) moving about over the tops of the rocks. This crustacean is closely related to the terrestrial woodlouse, and although larger is very similar in appearance. Despite being essentially a marine animal, it lives a virtually terrestrial existence. On hot days it loses a lot of water through evaporation and the blood becomes very concentrated. This, however, is probably a life-saving mechanism used by the animal to stay cool when temperatures approach the lethal limit. It works in the same way as sweating in mammals, through the cooling effect of evaporation, and enables the animal to maintain a temperature several degrees below that of the surrounding air. When water is once more available at high tide, it can be taken up by this animal via both the mouth and the anus.

On the lower shore, in places shaded from the direct rays of the sun, the beadlet anemone (*Actinia equina*) (**II, 9**) is commonly seen, although it remains closed when uncovered by the tide and looks like a blob of red jelly. When the tide comes in, or if it should be living in a pool, the tentacles will open out and its characteristic blue spots can be seen around the top of the column. When covered by water, the anemone feeds on any small animal life which may bump into the tentacles. These are armed with batteries of stinging cells which, at the slightest touch, shoot out long, penetrating barbed threads that pierce the prey and inject a paralytic poison. The tentacles then close inwards and thrust the captured animal into the mouth which lies in the middle of the flower-like disc.

Apart from these more conspicuous animals of the shore, there are many more which retreat beneath rocks and stones at low tide; they may be found by turning over the rocks (which should of course always be returned to their original position). The lower shore is often the best place to start searching. Many animals such as the crumb-of-bread sponge (*Halichondria panicea*) (**II, 28**) attach themselves to the under side of rocks. The coat-of-mail shell or grey chiton (*Lepidochitona cinereus*) (**I, 12**) uses its foot to cling on in the same way as the limpet, while the common saddle oyster (*Anomia ephippium*) (**II, 24**) actually cements itself to the rock with a lime-impregnated holdfast or byssus which extends through a notch in the shell near the hinge.

The sea-spiders *Nymphon gracile* (p. 81) and *Pycnogonum littorale* (**II, 33**) cling closely to rocks and may easily pass unnoticed. The males of these inconspicuous but interesting animals take care of the egg bundles which are abandoned by the females; they have a special pair of appendages with which to carry them.

The common shore crab (*Carcinus maenas*) (**II, 34**) is very common on all shores—not only where there are rocks, but also on sand and mud. This crab, in common with many other shore animals, shows peaks of activity which are timed to coincide with high tide and darkness. In summer months on shores where there are sufficient rocks or pools in which to shelter, they move up the shore with the advancing tide and then remain there when the tide ebbs. In the winter, however, most crabs move back down the shore again with the ebbing tide. In very cold weather no upshore migrations occur with the tide and it seems that most crabs move off into deeper water at such times.

Shore crabs are often found infected with a parasitic barnacle known as *Sacculina carcini*. When fully developed this parasite appears as a sac-like bulge between the thorax and the abdomen which prises the abdomen away from its usual position tucked neatly under the thorax. The bulge should not be confused with eggs which are carried in the same position by the female. The parasite has an astonishing life history. The small, female planktonic larval stage attaches to a host crab and assumes a sac-like form which develops a hollow dart-like structure. This is pushed into the crab and through it, the internal contents of the sac are injected into the body cavity. The mass of cells takes up position under the intestine and starts to grow, sending root-like structures out to all parts of the host's body. Soon it is large enough to burst out through the crab's shell between the thorax and the abdomen. The larger male planktonic larval stage may then inject itself into the female and so be in a position to fertilise her. Infected crabs lose their ability to reproduce and males tend to take on the secondary sexual characteristics of females and behave in the same way as females carrying eggs.

The hairy porcelain crab (*Porcellana platycheles*) (**II, 30**) has great flat nippers and a tiny body which it keeps close against the underside of rocks, to which it clings tightly with the sharp points on the ends of its legs. It is a suspension feeder and it creates its own feeding current by beating the water with a pair of hairy mouth parts.

Under boulders at the bottom of the shore on a good low spring tide, echinoderms may be found—brittle stars and, where deep gullies provide shelter, sea-cucumbers (described in the following sections).

Boulders too provide a habitat for sea-squirts like the solitary *Ciona intestinalis* (**I, 14**) or the beautiful blue-green or orange colonial 'star' species *Botryllus schlosseri* (**II, 10**).

Seaweed, lying thick and damp on the shore, provides an ideal habitat for many animals which would suffer from desiccation if not given the protection afforded by these plants. The flat periwinkle (*Littorina littoralis*) (**II, 20**) for instance is always associated with weed, particularly *Fucus* or *Ascophyllum*. This snail remains crawling around on the damp surface of the weed even when the tide is out and never needs to close the shell in the way the other winkles must to avoid losing water. It lays its eggs in small gelatinous clumps attached to the weed and the young hatch in the crawling stage, already placed in the adult environment. Other snails like the top shells *Gibbula cinerarid* (p. 70) and *Calliostoma zizyphinum* (**II, 18**) are usually found associated with weed, and the sea-lemon (*Archidoris pseudoargus*) (**I, 15**), which is a sea-slug, must keep to damp places because without the protection of a shell it would quickly suffer desiccation if exposed to the sun or the drying effects of wind. The sea-lemon may appear in some numbers on the shore in the spring when it comes to lay its egg masses; these appear as white, spirally coiled ribbons hanging from the walls of damp rock crevices or the underside of boulders.

A common animal which is often mistaken for weed by the casual observer is the polyzoan *Flustra foliacea* (**I, 17**), also called hornwrack. Its plant-like fronds may be found in large amounts on the shore particularly after a storm. Examination under a hand lens or low power microscope will reveal the structure of box-like zooids which comprise this colonial animal. If viewed under water the protracted tentacles are easily visible and clearly demonstrate the fact that this organism is indeed an animal.

There are a number of rock-boring molluscs which may be found on the shore. The common piddock (*Pholas dactylus*) (p. 77) and the oval piddock (*Zirfaea crispata*) (p. 77) are two which are probably the most commonly seen. They bore into the rock by mechanical action—their shells are provided with rows of spines near the front edge and, by a rocking motion of the mollusc in its burrow, these teeth slowly rasp away at the substratum and enlarge the hole. The oval piddock is usually found only in soft rock or clay while the common piddock also bores into wood, firm sand or peat. The common piddock, if seen at night, produces a greenish-blue luminescent light which emanates from a mucous secretion liberated into the exhalent siphon.

Of all the rocky shore habitats it is probably the rock pool which holds the greatest attraction for the naturalist. Here marine animals may be seen in their true environment. The dahlia anemone (*Tealia felina*) (**II, 27**) and sea-firs like *Tubularia indivisa* (**I, 18**) have their tentacles open, displaying their colours. Occasionally one may see a hermit crab in its disused snail shell sometimes with the 'parasitic' anemone (*Calliactis parasitica*) (**II, 12**) attached. This anemone is not really a parasite but a commensal—that is, the two animals have a mutually beneficial relationship. The anemone is carried around on the hermit crab's back and can share its meals as well as catching small animals which are disturbed when the crab walks across the bottom of the pool. In return, the anemone provides some protection for the crab as predators will be deterred by the stinging cells of its tentacles.

Pools provide a habitat for species such as the common prawn (*Palaemon serratus*) (**II, 25**) and the pugnacious fiddler or velvet swimming crab (*Macropipus puber*) which are usually only to be found below low tide mark. The echinoderms are very sensitive to desiccation and are rarely seen out of water, but careful searching in pools low on the shore may reveal the bright orange-red common starfish (*Asterias rubens*) (**II, 14**). This predator feeds largely on bivalve molluscs by humping itself over the body of its prey, gripping the shell with its tube feet, pulling the valves apart and everting its stomach into the shell to digest the soft parts within. The little cushion-star or starlet (*Asterina gibbosa*) (**II, 8**) is also a common inhabitant of rock pools, feeding on the dead particles of organic matter or detritus which lie on the bottom, or the diatoms and bacteria which coat the rocks. The purple-tipped green sea-urchin (*Psammechinus miliaris*) (**I, 16**) is sometimes common below the intertidal zone but may also be found in pools on the shore. It usually decorates itself with small pieces of weed, shell or stones, perhaps as a form of camouflage.

There are a number of fishes which inhabit the seashore. These are mostly confined to pools although the smallest puddle beneath a boulder may act as a refuge for tiny fish when the tide has retreated. There would be little chance of survival for fishes' eggs or larvae on the shore—the dangers of their being stranded at low water or washed out to sea would be too great. The shore fish have therefore abandoned their ancestral habit of shedding the gametes into the sea and leaving the eggs and young to fend for themselves, and they lay fewer eggs which are carefully nurtured. The female common blenny or shanny (*Blennius pholis*)(**II, 26**) lays her tiny, amber-coloured eggs on the top and sides of a rock crevice in spring or early summer and then abandons them, leaving the male to stand guard until they hatch. The eggs laid by this species number a few hundred, in contrast to the many thousands laid by species of fish in the open sea which broadcast their eggs and do not care for them. The gunnel or butterfish (*Pholis gunnellus*)(**II, 31**) also lays a relatively small number of eggs and the male and female both curl their long ribbon-like bodies round them, rolling them into a ball and protecting them until they hatch. The fifteen-spined stickleback (*Spinachia spinachia*)(**II, 13**) provides the most sophisticated care of all for its eggs and young. A mass of seaweeds is woven together and bound by a glutinous thread secreted by the kidneys. This nest is anchored to a growing plant and the female deposits her eggs inside. The male then guards them, and the fry when they hatch, until they are large enough to fend for themselves.

In the Baltic, several fish normally associated with freshwater rivers and lakes are found close to the shore where salinities are low. The perch (*Perca fluviatalis*) (p. 84) is common. It is a predator, feeding mainly on small crustaceans and young fish; it may be caught on a rod and line by spinning, or on a hook baited with a small fish, worm or insect. The zander or pike-perch (*Stizostedion lucioperca*) (p. 84) is related to the perch and bears some resemblance to it, but the body is more slender. It is less widely distributed than the perch but may be caught on rod and line with a spinner or baited hook. The zander prefers turbid water and so avoids competition with the pike, which tends to thrive in clearer water. When young these fish live in small shoals, but as they grow older they become more solitary. Large pike (*Esox lucius*)(p. 84) frequent deeper open water in the Baltic area but smaller specimens occur closer inshore and may be caught on rod and line with a spinner or a hook baited with a small dead fish.

A great many birds visit the rocky shore, particularly at low tide, when a wealth and variety of food is available. Most of these birds are only temporary visitors who may either move offshore or inland when the tide returns. There are also a number of birds which may be seen on the water close to shore. These are seen equally well from a boat or by a land-based observer and of course many will not be restricted to any particular type of shore and could be expected in rocky, sandy or muddy areas.

The grebes are a family of diving birds which characteristically have short legs set far back on the body; they have a short tail, short wings and the toes are lobed; in flight the head is held low and they appear weak and hurried. The grebes breed and spend the summer months inland in the vicinity of fresh water but during the winter they are often to be seen near the shore in sheltered areas and estuaries. The largest species is the great crested grebe (*Podiceps cristatus*) (p. 86) noted for its courtship display. They usually occur in small parties. The distinctive double crest is lost in winter and it may be confused with the red-necked grebe (*Podiceps grisegena*) (p. 86) although this has a shorter neck. The slavonian or horned grebe (*Podiceps auritus*) (p. 86) is an uncommon bird throughout north-west Europe, though more widespread on the coasts in winter than during the summer when it is restricted to northern areas.

The heron or grey heron (*Ardea cinerea*) (p. 86) is a common bird of the shore and estuary as well as on freshwater lakes and rivers inland. It hunts its prey at the edge of the water by stalking or standing absolutely still and striking rapidly with its long yellow bill. It usually takes small fish, and eels also seem to be an important part of its diet; but it will catch insects, amphibians, reptiles, young birds and small mammals as well.

A variety of ducks are to be seen near the shore. The mallard (*Anas platyrhynchos*) (p. 86) is mainly vegetarian when inland on fresh water, but when on the coast it will also feed on small snails, prawns, shrimps, crabs and similar crustaceans. The tufted duck (*Aythya fuligula*) (p. 88) normally remains on fresh water during the breeding season but at other times may be seen on the seashore. It is a diving species and may descend to depths of 2 metres to take water plants and to a lesser extent small fish and crustaceans. Other ducks which dive for their food are the goldeneye (*Bucephala clangula*) (p. 88) and the scoters (described in Chapter Three). These take small shellfish and are to be seen on the coast during the winter.

The mute swan (*Cygnus olor*) (p. 88) is present on our coasts all the year round, but the other swans of north-west Europe, the whooper (*Cygnus cygnus*) (p. 88) and the Bewick's (*Cygnus columbianus*) (p. 88), together with the bean goose (*Anser fabalis*) (p. 88), are winter visitors from the far north where they breed and pass the summer months. These birds all feed on vegetation. Both swans and geese browse on the shore but the swans also dip their necks below the surface to take plant material from the bottom.

Although hardly a characteristic bird of the shore, the black grouse (*Lyrurus tetrix*) (p. 88) may be seen on some of the Baltic islands, where they may occasionally nest. The males or blackcocks perform spectacular displays at special sites called 'leks'.

Unlike most of the wading birds which often prefer sandy or muddy shores, the turnstone (*Arenaria interpres*) (p. 88) is usually found on rocky shores, although it is also to be seen on stony beaches and, rarely, inland near fresh water. It is a winter visitor to north-west Europe and it nests from the Baltic northwards. As the name suggests, turnstones pick over stones, seaweed and other objects on the shore in search of sandhoppers, insects and small fish. They are usually seen singly or in small groups and sometimes with other small shore birds.

The hooded or grey crow (*Corvus corone cornix*) (p. 89) is another species which, although not strictly a shore bird, is a frequent visitor to the coast especially where there are rocky cliffs. It is closely related to the carrion crow (*Corvus corone corone*) and like that bird has a very catholic diet, taking insects, grain, other birds' eggs and young, carrion, fruit and fish.

The white wagtail (*Motacilla alba alba*) (p. 92) on mainland Europe and the pied wagtail (*Motacilla alba yarrellii*) (p. 92) in the British Isles are also by no means confined to the seashore but individuals often make a successful living there. They may be seen running over the rocks in pursuit of small insects, sometimes in the company of the rock pipit (*Anthus spinoletta*) (II, 17). This little bird also takes insects but its main food is small molluscs which it picks up from cracks and crevices in the rock.

There are a great many more birds which are likely to be seen on the coast. Some of these are characteristic of sandy and muddy shores and are dealt with in the relevant sections which follow in this chapter; and others, like the gulls and terns, which also occur in the open sea are covered in Chapter Three. Of course many birds of the open sea can be seen by the land-based observer, particularly during the nesting season. It is often the rocky shores and cliffs of coasts that oceanic birds like the auks, gannets, petrels and shearwaters choose to rear their young.

In isolated areas—the remote islands and inlets of Scotland and the west of Ireland—seals haul themselves ashore and may be seen basking in the sun. These marine mammals are active hunters of fish and they can dive and stay underwater while hunting for up to fifteen minutes at a time. Unlike the sealions, their back flippers are virtually useless on land as they cannot be turned forwards, and so the seals move rather clumsily when out of the water. The grey seal (*Halichoerus grypus*) (p. 92) comes ashore for the breeding season and the young are born on rocky ledges or isolated beaches well away from the sea. For the first two weeks they are unable to swim; they stay away from the sea as they would probably drown if they fell in. The common seal (*Phoca vitulina*) (p. 92) on the other hand gives birth to pups which are able to swim almost immediately. The mother may land on a remote sand- or mudbank and give birth to a pup which is able to swim away with her by the next tide.

THE SANDY SHORE

Sandy shores occur wherever conditions are suitable for deposition, in the small sheltered embayments between rocky headlands or over long distances of coastline such as along many shores of the English Channel, the North Sea and along the western coast of France. They are built up where the predominant waves are of the constructive type. These waves, although pounding the beach with some considerable force, tend to build up material because the swash onshore is of greater intensity than the backwash. Destructive waves, which remove material from the beach, occur when the frequency of the waves is reduced so that the backwash from one wave reduces the force of wash of the next thus reducing its effect in moving material onshore. Destructive waves are often the result of localised wind conditions.

Sandy shores sometimes become unstable. They are usually in a dynamic state with material being removed by undertow and longshore currents at the same time, and as fast, as it is being deposited. If the equilibrium is upset, for example by dredging offshore, a sandy beach can be destroyed very quickly. Even without intervention from man a storm can change the composition of a beach overnight. At seaside resorts where sand is at a premium considerable efforts are made to stabilise beaches by building breakwaters and groynes.

Material which has been carried off the beach is often deposited offshore and may form a bar. The yachtsman approaching a sandy beach would be well advised to remember this fact as these bars can change depth and position very quickly and charts cannot be relied upon to give more than a general warning as to the likelihood of their presence. Sand bars of this kind often break the surface, either in a continuous line or as a series of small sandy islands which may disappear at high tide. To go aground on one of these on a falling tide could well result in a long wait in an uncomfortable position.

Sand grains are usually composed of quartz which has been removed from the parent rock by the processes of erosion. On the shore they are usually surrounded by a cushion-like coating of water which prevents abrasion so they retain their sharp angular shape. Desert sands on the other hand are dry and so become rounded and smooth as a result of wind action. Larger stones and pebbles on the shore become rounded in a similar way due to wave action, but as shores composed of these support little or no life they will not be considered further here. Pebble shores are usually steeper than sandy shores which in turn are steeper than muddy shores which tend to be almost flat. Sands usually contain large quantities of broken shell particles which, because of their soft and soluble nature, are rapidly worn away—but of course these are continually replaced from freshly dead mollusc shells which quickly become broken up.

Where the land lying behind sandy beaches is reasonably flat, onshore winds blowing dry sand inland from the top of the shore initiate the first step in the formation of sand dunes. Dunes are present for long distances on coasts such as those of North Wales and Lancashire in Britain, the Belgian, Dutch, German and Danish coasts and that of western France. Although sandy shores support very little growth of marine plants the vegetation of the dune complex behind the shore is of considerable interest. Often the first plants to colonise the sand above high water mark occur in a narrow band at the level of the equinoctial spring tides. Here piles of seaweed and other organic material are deposited and when covered by wind-blown sand are an ideal site for the germination in spring of annuals such as saltwort (*Salsola kali*) (**III, 4**) and the sea-rocket (*Cakile maritima*) (**III, 1**).

Once these species have grown they act as a small windbreak trapping more sand and so starting the development of a dune. Permanent sand dunes develop where these embryonic dunes are further colonised by perennial species such as sea-couchgrass (*Agropyron junceiforme*) (**III, 18, 18a**) and sea-sandwort (*Honkenya peploides*) (**IV, 3**). The creeping network of stems and shoots produced by these plants prevents wind erosion and as further sand is deposited they grow up through the sand and so the development of the dune proceeds. As the dune develops (and this process can be followed by walking back from the shore to the more mature dunes) further plant species take over. Marram grass (*Ammophila arenaria*) (**III, 2**) appears as soon as the dune is high enough to keep the roots out of the salt water of the highest tides. This species grows vigorously, producing numerous side shoots, and quickly grows up through any covering of sand it may receive in windy weather. In this way marram grass dunes may reach a height of 50 metres.

Behind the first ridge of the sand dunes a number of flowering plants start to appear: the yellow-flowered ragwort (*Senecio jacobaea*) (**IV, 8**) and the sea-spurge (*Euphorbia paralias*) (**IV, 1**) with its thick fleshy leaves are common. The early forget-me-not (*Myosotis ramosissima*) (**IV, 4**) is sometimes to be found further back from the sea. The English stonecrop (*Sedum anglicum*) (**IV, 5**) is a dwarf plant with white petals which are pink underneath; it forms spreading mats over the sand. The sea-bindweed (*Calystegia soldanella*) (**IV, 9**) with its pink trumpet-shaped flowers puts down deep roots into the sand in the upper dunes. The annual sea-blite (*Suaeda maritima*) (**III, 3**) spreads over the sand, and the prickly sea-holly (*Eryngium maritimum*) (**IV, 7**) used to be a common plant but today is not usual outside nature reserves. Further back from the sea one can expect to find grasses such as the sand cat's-tail (*Phleum arenarium*) (**III, 5**) and a variety of mosses and lichens.

Sand dunes are a rather unfavourable environment for small animals as there is very little vegetation cover, the sand contains little organic matter and is very dry. It warms up very quickly when the sun shines but it soon cools at night or on cloudy days. Although they are outside the scope of this book, it should be mentioned that those animals which do manage to colonise the dunes fall into two categories. The first includes a number of species characteristic of grasslands, which survive in the dune environment, such as various insects, rabbits, rats, mice, stoats, weasels, snakes and lizards. The second category contains various duneland specialists like the sand-hill snail or pointed snail (*Cochlicella acuta*) (p. 73), the greater dune robber-fly (*Philonicus albiceps*) (p. 81) and numerous other small insects. Many of the sand dune animals are nocturnal, remaining among the base of marram clumps or buried in the sand during the day to avoid the desiccating effects of the sun.

When the tide recedes on a sandy shore there is generally little sign of life apart from some worm casts, perhaps, and a few broken shells. There is however a wealth of living creatures to be found burrowing beneath the surface and these can be found either by digging or sieving the sand; a plastic gravy strainer is useful for this job.

As there are few seaweeds to be found on sandy shores many of the animals depend on the plant life of the phytoplankton for their primary source of food and so one can expect to find a number of species which are adapted for suspension feeding when the tide is in. Suspension feeding has evolved independently in several different groups of animals, and the mechanisms which have developed, to pump and filter the tiny organic particles and phytoplankton from the water, have various origins. The simplest suspension feeders are the sponges which are generally scarce on sandy shores although the boring sponge (*Cliona celata*) (**IV, 17**) may be present in the shells of bivalves.

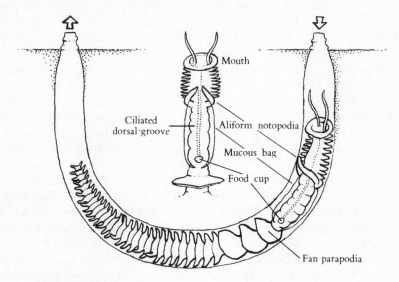

Fig. 2. Chaetopterus feeding.

The parchment worm (*Chaetopterus variopedatus*) (**IV, 12**) can be collected by careful digging. It has developed a novel method of feeding which is worth describing. The animal lives in a U-shaped, parchment-like tube and has a highly modified body structure: on the twelfth segment there are two extended wing-like processes (projections) and the fourteenth, fifteenth and sixteenth segments are called fans. These fans resemble piston rings and fit closely to the sides of the parchment tube (see Fig. 2). The beating of the fans creates a water current which passes from the anterior to the posterior end of the tube, forcing the water to pass through a bag-like mucus net which the processes on the twelfth segment secrete. The organic particles and plankton in the water current are trapped in the net which at regular intervals is rolled up into a ball and transported to the mouth where it is swallowed; a new net is then secreted and the process repeated.

One of the rarer suspension-feeding crustaceans which may be found in the sand on an exceptionally low spring tide is the burrowing prawn (*Upogebia deltaura*)(**IV, 11**). This species, like the hairy porcelain crab, to which it is closely related, strains water through its hairy mouth parts. It lives in very deep burrows with several openings to the surface and creates a water current through the burrow with its modified, hinder swimming appendages.

The molluscs are probably the best-represented phyla among the suspension feeders to be found on the sandy shore. In certain areas the slipper limpet (*Crepidula fornicata*) (**III, 11**) may be extremely common, living on the sand surface usually in chains of individuals, with the oldest and largest at the bottom and often attached to a stone or boulder. Young animals at the top of the chain are always males and these change sex as they age. Feeding currents are created by the gills through a slight gap left between the shell and the substratum, often the shell of the animal beneath, and food particles are trapped on mucus covering the gill surface; this is then transported to the mouth where it is consumed. The slipper limpet was introduced to Europe from America with imported oysters, probably in about 1880, and on several occasions since. It is a severe pest on oyster beds where it competes for space by settling on the surface of clean shell on the seabed which is intended for oyster spat. The presence of slipper limpets in certain areas of Britain has led to the Ministry for Agriculture, Fisheries and Food banning the transport of oysters and other molluscan shellfish from designated areas unless it has been ascertained that there is no danger of spreading this pest. Oyster catches from infected areas can be rendered safe if necessary by dipping them in a strong brine solution which kills any slipper limpets that might be present. Yachtsmen and naturalists collecting specimens in infected areas should take great care not to discard samples in places where slipper limpets might not be present as this would risk the danger of introduction.

The many suspension-feeding bivalves which live in sand use their gills to create water currents through the mantle cavity, to obtain suspended food and oxygen. Particles are trapped in mucus on the surface of the gills and carried to the mouth by special cilia. If there is a high level of particulate matter in the water and the bivalve cannot cope with it all, a certain amount of material is rejected as pseudofaeces. Most bivalves such as the common cockle (*Cerastoderma edule*)(**III, 10**), the various venus clams and the razor shells—*Ensis ensis* (**III, 25**) and *E. siliqua* (p. 77)—burrow only to shallow depths, just below the surface of the sand, and possess short siphons which project up to the water to draw in water for feeding and respiration and to expel waste products. At low tide, the razor clams use their long powerful foot to dig deep down away from the dangers of predation and it is very difficult to catch them, but a search by torchlight on a bright moonlit night, at low water spring tide, may reveal large numbers of these animals, all extending the top of the shell above the surface. Care must be taken in approaching, as the slightest disturbance of the surface of the sand by a clumsy footfall may cause the animal to disappear below with all speed—and attempts to dig it out are inevitably futile.

While suspension feeding provides a means for many animals to make a living on the shore, there are others which derive their nutriment from the sand itself; as a general group they are termed the deposit feeders. Of course quartz grains alone have no food value whatsoever, but all sands contain a certain amount of dead, particulate, organic material derived from decaying seaweed, perhaps brought in by currents from elsewhere, from faeces, from the dead remains of animals and from organic material blown onto the shore from the land behind. Organic particles are very light so they are easily picked up and carried in the water if there is any turbulence. Sandy beaches in calm areas therefore tend to have more organic matter than in rough places and this is reflected in the numbers of animals living there. As water trickles down through the sand it gradually becomes deoxygenated due to the respiratory activity of micro-organisms which feed on the organic matter and also because of a certain amount of chemical oxidation of substances in the sand. The level in the sand where all the oxygen has been used up—the anaerobic or sulphide layer as it is known—is very obvious from its black appearance, due to the presence of ferrous sulphide, and the smell of hydrogen sulphide which emanates if it is disturbed.

The common lugworm (*Arenicola marina*) (**III, 13**) lives in a mucus-lined burrow, usually in the sulphide layer; this burrow has a tail shaft which extends to the sand surface. The worm is a deposit feeder and as it ingests the sand at the head of the burrow, this is replenished with more that falls down from above, so causing a depression on the sand's surface. From time to time the tail is extended up the shaft to the top of the burrow to defecate, producing the worm-like casts so widespread on a sandy beach at low tide. Water for respiratory purposes is drawn in through the tail shaft at high tide when the mouth of the burrow is covered, but at low tide the oxygen concentration of the water in the burrow falls to such a low level that the worm has to draw bubbles of air down from the surface to bathe the gills. If all the water in the burrow should drain away, the lugworm is able to respire atmospheric air, provided the body surface remains wet.

Often the only sign of life on a sandy beach is the tube of the sand mason (*Lanice conchilega*)(**III, 17**). This worm lives towards the lower shore and the tube is built of sand grains and shell fragments; it projects from the sand as a ragged tuft. At high tide the head of the worm moves to the top of the tube and long extensile tentacles explore a wide area of the sand's surface for food and building materials. When building the tube, sand and shell particles are taken into the mouth, where they are coated with a quick-setting cement and then placed into position where they immediately stick firmly into place. If danger threatens, the tentacles snap back into the tube and disappear. Although the sand mason has been considered for a long time to be simply a deposit feeder, it is now known that some of the tentacles are also used for suspension feeding, often at the same time as others are deposit feeding. Where there is a current, the upstream tentacles pick material up off the bottom and the downstream ones collect particles out of suspension. This behaviour is probably valuable where there is not much food on the bottom—for example in areas where there is a high population density.

There are a variety of molluscs which are deposit feeders. The beautiful pelican's foot shell (*Apporhais pespelicani*)(III, 12) digs into the sand and uses the siphon to project up to the surface to pick up organic matter like a vacuum cleaner. When a particular area has been exhausted this snail moves on to an adjacent site. There are also a number of deposit-feeding bivalves like the thin tellin (*Tellina tenuis*) (III, 20), the banded wedge shell (*Donax vittatus*)(III, 19) and the large sunset shell (*Gari depressa*) (IV, 15), which remain buried in the sand and extend long siphons to the surface to suck food off the bottom. These bivalves almost certainly use their siphons to suspension-feed as well when the tide is in.

In general those bivalve species inhabiting sandy shores have a thinner shape than those on muddy shores. Exceptions to this rule do exist (the cockle, for example) but they are far slower in burying themselves if disturbed than the narrow forms like *Tellina tenuis* or *Donax vittatus*. This is because sand is generally harder than mud and creates more resistance to an animal moving through it. Shore deposits range in their physical characteristics from a dilatent state—the condition of sand when it whitens and hardens under foot, to a thixotropic state—when pressure causes the surface to soften. The degree of dilatancy or thixotropy depends upon the particle size of the deposit and the amount of water held by it. Dilatent deposits have a large particle size and lower water content; while thixotropic deposits generally have a large percentage of small particles and a high water content.

Bivalves burrow by means of the foot which in most species is a flattened organ well designed for penetrating the deposit in which it lives. The sequence of events when burrowing is similar in many species. The foot is pushed into the deposit and the siphons close; water is then ejected from within the shell by a closing movement of the valves, the siphons open, the bottom end of the foot swells to create an anchor and the animal pulls itself downwards. The shells then gape open to fix the animal in position while the foot probes downwards for the next cycle. Striations and ribs on the shell surface may aid in holding the animal in the sand while this is happening and ejection of water from within the shell, together with a rocking motion executed by some species like cockles, probably tends to create a thixotropic condition in the deposit. This may happen in the same way as when the shore walker 'dabbles' wet sand with his foot in order to make it soft. If so, this would lessen resistance to the bivalve trying to move through it. The force with which bivalves are able to pull themselves downwards is often quite considerable—the razor clam, which has an advanced type of rounded foot, can exert a pull of 800 g downwards. This is exceptional and the figure is usually much less—but many species, when placed on the sand surface, are able to disappear with astonishing speed.

Several species of echinoderm which one is likely to find on sandy shores are also deposit feeders. The sea-potato (*Echinocardium cordatum*)(III, 15) shovels sand

Plate I

PLANTS AND ANIMALS OF ROCKY SHORES

1. *Cochlearia officinalis*: Scurvy grass
2. *Spergularia rupicola*: Rock sea-spurrey
3. *Xanthoria parietina*, var. *ectanea*: Orange lichen
4. *Verrucaria maura*: Rock-encrusting lichen
5. *Verrucaria mucosa*: Lichen of tidal rocks
6. *Limonium binervosum*: Rock sea-lavender
7. *Silene maritima*: Sea-campion
8. *Crithmum maritimum*: Samphire or rock samphire
9. *Brassica oleracea*: Wild cabbage

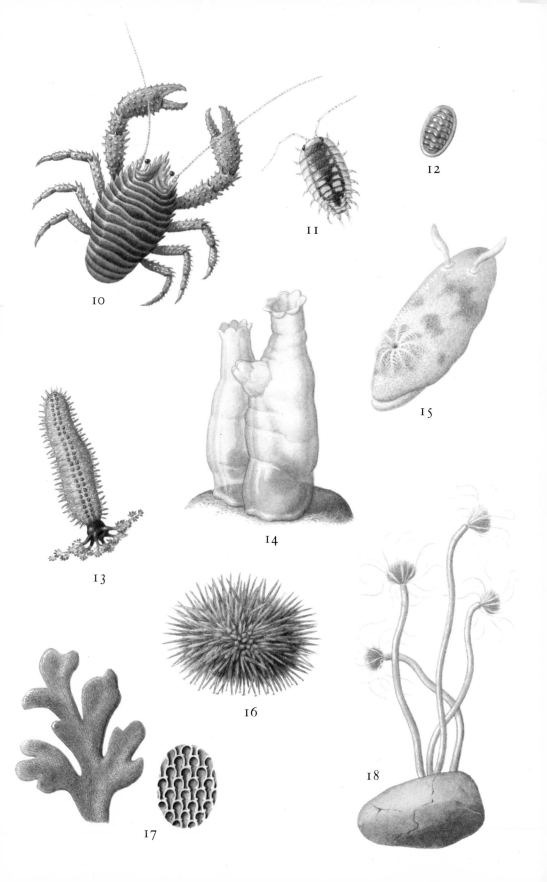

Plate I

PLANTS AND ANIMALS OF ROCKY SHORES

10. *Galathea squamifera*: Squat lobster
11. *Ligia oceanica*: Sea-slater
12. *Lepidochitona cinereus*: Grey chiton or coat-of-mail shell
13. *Cucumaria normani*: Sea-cucumber
14. *Ciona intestinalis*: Sea-squirt
15. *Archidoris pseudoargus*: Sea-lemon
16. *Psammechinus miliaris*: Green sea-urchin
17. *Flustra foliacea*: Hornwrack (left), with enlargement of structure detail
18. *Tubularia indivisa*: Sea-fir

Plate II

LIFE AROUND A ROCK POOL

1. *Armeria maritima*: Sea-pink or thrift
2. *Fucus serratus* (Serrated or toothed wrack) with *Spirorbis borealis* (tubeworm)
3. *Ascophyllum nodosum*: Knotted or egg wrack
4. *Laminaria digitata*: Oarweed
5. *Fucus vesiculosus*: Bladder wrack
6. *Pelvitia canaliculata*: Channelled wrack
7. *Rhodymenia palmata*: Dulse
8. *Asterina gibbosa*: Cushion-star or starlet
9. *Actinia equina*: Beadlet anemone
10. *Botryllus schlosseri*: Star sea-squirt
11. *Corallina officinalis*: Coral weed
12. *Eupagurus bernhardus* (hermit crab) in *Buccinum undatum* (whelk shell) carrying *Calliactis parasitica* ('parasitic' anemone)
13. *Spinachia spinachia*: 15-spined stickleback
14. *Asterias rubens*: Common starfish
15. *Enteromorpha intestinalis*: Grass kelp

16

17

18

19

20

21

22

23

24

25

26

27

28

29

30

31

32

33

34

Plate II

LIFE AROUND A ROCK POOL

16. *Nucella lapillus*: Common dogwhelk
17. *Anthus spinoletta*: Rock pipit
18. *Calliostoma zizyphinum*: Painted topshell
19. *Patella vulgata*: Common limpet
20. *Littorina littoralis*: Flat periwinkle
21. *Littorina littorea*: Edible or common periwinkle
22. *Balanus balanoides*: Acorn barnacle
23. *Mytilus edulis*: Common mussel
24. *Anomia ephippium*: Common saddle oyster
25. *Palaemon serratus*: Common prawn
26. *Blennius pholis*: Common blenny or shanny
27. *Tealia felina*: Dahlia anemone
28. *Halichondria panicea*: Crumb-of-bread sponge
29. *Ulva lactuca*: Sea-lettuce
30. *Porcellana platycheles*: Hairy porcelain crab
31. *Pholis gunnellus*: Gunnel or butterfish
32. *Lithophyllum incrustans*: Red encrusting seaweed
33. *Pycnogonum littorale*: Sea-spider
34. *Carcinus maenas*: Common shore crab

from beneath its body by means of specialised, flattened spines and digs itself in to a depth of several centimetres, where it lives in a chamber connected to the surface by a narrow passage. Secretions produced by the animal cement the sides of the passage to prevent collapse and through it, long tube feet are extended to the sand surface where they collect food. The worm-cucumber (*Leptosynapta inhaerens*) (**IV, 13**) burrows into the sand, leaving its tentacles projecting above the surface. The brittle star *Ophiura texturata* (**IV, 14**) burrows to shallow depths on the lower shore; it uses its suckerless tube feet to collect small organic particles, while large food particles are raked into the mouth by the arms.

Predators on the sandy shore include the sea-anemones, such as *Peachia hastata* (**III, 31**). This species has evolved a worm-like form and it burrows in the sand leaving just the tentacles spread out on the surface to trap minute zooplankton and other small prey with their stinging cells. The cat-worm (*Nephtys hombergi*) (**III, 16**) is a voracious predator with powerful jaws which are used to capture its prey, and the unusual-looking *Priapulus caudatus* (**IV, 10**) employs its eversible proboscis to capture slow-moving soft-bodied animals such as worms. The burrowing starfish (*Astropecten irregularis*) (**III, 26**) moves through the sand in search of bivalves, crustaceans and worms; it eats its prey whole and later disgorges the skeleton.

An alternative way to make a living on the shore is to be a scavenger, feeding on dead animal and plant material in whatever form it is available. Many crustaceans live a scavenger's existence. The common shrimp (*Crangon vulgaris*) (**III, 30**) is nocturnal— during the day it buries itself in the sand and although it may not be completely covered, its brown coloration prevents its detection by predators. A push net disturbs the shrimp and catches it as it leaves the safety of the sand. The masked crab (*Corystes cassivelaunus*) (**III, 28**) also burrows during the daytime and uses its very long antennae to create a tube to the surface through which it draws its respiratory water current. The amphipod crustacean *Talitris saltator* (**III, 29**) lives under rotting seaweed at the top of the shore during the daytime and emerges from its damp shelter at night. A trip to the beach with a torch after dark will reveal thousands of these sand-hoppers searching the upper shore for any organic material left by the tide during the day.

Sandy shores provide a rich feeding ground for a number of birds. Many of these are waders which feed on the infauna—that is, animals living below the surface of the sand. The exact type of food each species feeds on depends upon the type and length of bill possessed (see Fig. 3), the wide variety of bill type ensuring that there is little overlap between the food resource of one species of bird and another. The common ringed and kentish plovers—*Charadrius hiaticula* (**III, 21**) and *C. alexandrinus* (p. 88)—thus take surface-living animals as may the oystercatcher (*Haematopus ostralegus*) (**III, 23**) although this bird will also dig shallow-living shellfish, like cockles, out of the sand. These are broken open with hammer blows from the closed bill, but if a bivalve such as a mussel is slightly open in shallow water, it may be opened by stabbing the bill between the valves and into the soft parts within. Birds with slightly longer bills such as the redshank

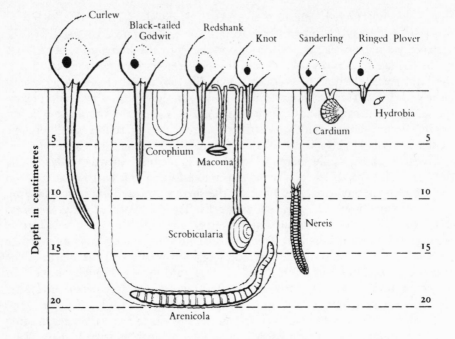

Fig.3. Length of bill of some common waders.

and greenshank (*Tringa totanus* (**VI, 17**) and *T. nebularia*) (p. 89) can take deeper-living bivalves and crustaceans. There are probably few species which escape from the curlew (*Numenius arquata*) (**VI, 4**) whose bill may be up to 150 mm in length.

The common shelduck (*Tadorna tadorna*) (**III, 22**) is a very conspicuous bird often to be seen on the shore feeding largely on surface-living molluscs and crustaceans, but it will also take worms and a small amount of plant material like *Enteromorpha*. The shelduck nests in sand dunes at the back of the shore and afterwards migrates to moulting grounds such as at the mouths of the Rhine and the Elbe. At these places huge numbers of birds assemble and moult, and during this time they temporarily lose the ability to fly.

Sand dunes also provide a habitat for seed- and insect-eating birds. The snow bunting (*Plectrophenax nivalis*) nests in Scandinavia and in northern Scotland, but in the winter it moves south to coastal areas of northern Europe and may often be seen feeding on the seeds of grass, rushes and weeds behind the shore.

THE MUDDY SHORE

Wherever calm sheltered conditions occur, or where the strength of tidal currents is reduced, finely divided particles are deposited and give rise to muddy shores.

Such conditions prevail in many of our estuaries and sheltered sea lochs so often frequented by yachtsmen. These places not only provide shelter from wind and waves and an excellent holding ground for anchoring but wonderful opportunities for the naturalist.

Muddy shores differ from sandy shores in a number of respects. Physically they are usually much flatter, hence the term mudflat, and for this reason they often extend over large areas when the tide recedes. Being flat, muddy shores do not drain as effectively as other types and so a large amount of water remains within the deposit. This effect is enhanced by the capillary action of the closely packed particles. The particles themselves are smaller, although there is usually a small amount of fine sand mixed with mud as well as a far higher organic content which consists of small particles of detritus in the process of being broken down by bacteria. Because of the higher organic component of muds and their low permeability, the sulphide layer, mentioned in the section on sandy shores, is much closer to the surface than in sand. Walking over a mudflat, one's feet usually sink well down, breaking into the sulphide layer and so releasing the characteristic odour of hydrogen sulphide.

The muddy shores of estuaries sheltered from wave action are covered by a very distinctive type of vegetation termed saltmarsh. This vegetation extends to below the level of high water but at low water shallow-sloping mudbanks may extend for long distances and be entirely free of plant cover, except perhaps for a few algae. The first land plants to colonise the shore are often species of glasswort such as *Salicornia europaea* (**VI, 18**); a succulent plant, its stem swollen with water storage tissue, it may form dense meadows in the summer months. Glassworts are annual plants which seed in autumn and begin to grow in spring; in winter all that remains are a few tough, woody stems. Further away from the upper limits of the tide a variety of flowering species occurs: sea-spurrey (*Spergularia marina*), perennial sea-spurrey (*Spergularia media*) (**V, 2**), sea-lavender (*Limonium vulgaris*) (**V, 5**), the sea-milkwort or black saltwort (*Glaux maritima*) (**V, 1**) which forms spreading mats with its slender, creeping form, and sea-purslane (*Halimione portulacoides*) (**VI, 15**) which is a low-growing shrubby member of the spinach family. Further up still, in the region of the strandline, the marsh is usually dominated by the sea-rush (*Juncus maritimus*) (**V, 3**). This may cover extensive areas where conditions are right but if there is any influx of freshwater streams at this level a freshwater swamp community may develop, dominated by the common reed (*Phragmites communis*) (p. 67). Behind the saltmarsh above the highest level of the tide a variety of coast-loving grasses such as the annual beard grass (*Polypogon monspeliensis*) (**V, 7**) starts to appear.

In several estuaries of north-west Europe extensive areas of saltmarsh are dominated by rice grass or Townsend's cord-grass (*Spartina × townsendii*) (**VI, 14**). The species arose as a fertile hybrid through the natural crossing of a European species (*S. maritima*) and an American species (*S. alterniflora*) which was probably introduced by accident to the Southampton area, on the south coast of England, in the nineteenth century. It is unusual for a hybrid to be fertile but some strains of this species are. It also shows the phenomenon of hybrid vigour, being an aggressive form which has made extensive invasions into estuarine mudflats. It has been introduced into a number of areas particularly in Holland to aid reclamation of tidal lands. Unlike the native *Spartina*, it can act as the primary coloniser of bare mud; it spreads fast and its matted roots quickly cause the level of the mud to build up, converting wet areas to dry meadow land.

In the last fifty years the spread of *Spartina* has been arrested by a disease known as 'die back' in which the underground buds die and the rhizome apices rot. The disease hits the plant either along the margins of channels or in low-lying areas in the centre of the marsh. It seems that although more tolerant of wet areas than many other saltmarsh plants, the hybrid *Spartina* is not completely resistant. There is no evidence that the disease is caused by pollutants or pathogens.

On open shores where more marine conditions exist, and it is sheltered and muddy, beds of eel-grass or common grass wrack (*Zostera marina*) (**VI, 7**) and the smaller *Zostera nana* may sometimes be found. These beds are unfortunately less extensive now than they were in the past because of a wasting disease which spread rapidly and wiped out many beds in Europe and North America in the 1930s. Even today there is little sign of the return of this plant in many of the areas where it was present in the past. The various species of eel-grass which occur worldwide along temperate coasts are the only truly marine flowering plants found in north-west Europe. Certain animals such as the cotton-spinner or sea-cucumber (*Holothuria forskali*) (**VI, 5**) and some species of anemone (which will be mentioned later) are often found in association with beds of eel-grass.

The highly productive saltmarsh communities on the borders of estuaries give rise to a great deal of rotting vegetation which, due to physical, chemical and biological factors, becomes finely divided to produce the fine, wet, organic dust or detritus which is carried in suspension and deposited on the bottom and sides of estuaries over wide areas. The detritus, combined with large quantities of live phytoplankton growing in the nutrient-rich waters of the estuary, provides a rich food source for animal life. Thus, although the numbers of species of animals in estuaries are low and confined to those which are adapted to the special conditions of fluctuating salinity, the biomass and productivity of the animal populations tend to be high.

Living on the surface of the mud of estuaries at low tide, the tiny snail *Hydrobia ulvae* (**VI, 24**) occurs in huge numbers, literally millions, and when the tide returns it may float to the surface on rafts of its own mucus, so becoming a temporary member of the plankton. *Hydrobia* is a deposit feeder and it seems likely that an important part of its diet is the bacterial film which coats the organic particles that it eats. The mucus raft also serves as a trap for diatoms and other small suspended particles which are eaten by the snail. *Hydrobia* breeds over several months during the summer, laying up to about 25 eggs enclosed in a gelatinous capsule. These are usually attached to the shell of another *Hydrobia*. When they hatch there is a very short larval stage and then the young snail settles to the mud surface. The short larval stage is probably an adaptation to estuarine life which avoids the danger of larvae being carried out to sea and away from suitable areas for the adult snail.

There are rarely many signs of life on the mudflat but the presence of the peppery furrow shell (*Scrobicularia plana*) (**VI, 20**) is often marked in muddy sand by the presence of star-shaped patterns on the surface. This bivalve is a deposit feeder which uses its long inhalent siphon to suck up particles from the top of the mud; the animal lives in a burrow, usually at a depth of about 14 cm, and the siphon is extended up and out in directions which alter from time to time until a full circle has been completed. It is this behaviour which creates the patterns, which may be very numerous—population densities of up to 1000 per square metre have been recorded.

Another deposit-feeding bivalve, which occurs in huge numbers on muddy shores, is the Baltic tellin (*Macoma balthica*) (**VI, 19**). Numbers of 5000 per square metre have been recorded and there appears to be competition for space between *Macoma* and *Scrobicularia*. Although the two species may occur together on the same shore, where there are high densities of one, the other will only be present in small numbers. *Macoma* can quickly clear the surface of mud within reach of its inhalent siphon and then it moves on. The tracks made in the mud as the animal progresses are often clearly visible and the fact that some of them are U-shaped and orientated towards or away from the sun suggests that this bivalve uses similar behaviour patterns to those described for the periwinkle on the rocky shore for maintaining its position on the mudflat.

The small amphipod *Corophium volutator* (p. 77) is another very common animal of estuaries. It occurs in very high densities and the presence of these crustaceans is apparent from large numbers of small holes in the mud which are the entrances to their U-shaped burrows. The large specimens are the only ones that are likely to be seen as they frequently leave their burrows and crawl around on the mud, using their long antennae to pull themselves and their rear appendages to push themselves along, in a kind of looping movement. *Corophium* appears to be sensitive to hydrostatic pressure, so when the tide is flooding and hydrostatic pressure is increasing it leaves its burrow and walks around on the surface of the mud. When pressure decreases this stimulates it to start swimming. On leaving the burrow on the flooding tide *Corophium* thus tends to be carried up the beach, and by starting to swim when the pressure decreases it is carried back down the beach again with the ebbing tide towards the place where it started. Adaptations of this kind are apparently present in other shore animals such as the sea-spiders and various mysids and they appear to have arisen as a means by which the animals are able to return to their original position on the shore.

The butterfish or cross-cut carpet shell (*Venerupis decussata*) (**VI, 30**) likes muddy areas and makes very good eating; in France it is sought after as much as or more than the common European or flat oyster (*Ostrea edulis*) (**VI, 29**). This species is also often found in and near estuaries in places where the substratum is firmer, and where there is shell or gravel on the bottom. The American hard-shell clam (*Mercenaria mercenaria*) or quahog (p. 73) was introduced to Europe from the east coast of the United States, probably via the kitchens of Atlantic liners, and it now occurs in the vicinity of warm water outfalls from power stations and on commercial beds. The gaper shell (*Mya arenaria*) (**VI, 31**) gets its name because if dug out of the mud it is quite incapable of withdrawing the enormously enlarged siphons, and the shell gapes open with the large fleshy mass of the siphons protruding. This clam grows as it burrows into the mud so the whole tunnel becomes filled with the body, which may be up to 30 cm in length. The foot is much reduced in animals of this size and if dug up they find it very difficult to burrow again. Where oysters and other bivalves are present on the shore the oyster drill or sting winkle (*Ocenebra erinacea*) (**VI, 26**) is almost certain to be found. The neat little tell-tale holes drilled by the rasping tooth-like radula of this predatory snail will be apparent among some of the dead shell material even if the snail itself cannot be located.

The muddy shores in fully marine situations are colonised by a wider range of species than in estuaries. On rocks and stones among the mud one can expect to find the orange or scarlet sponge *Hymeniacidon sanguinea* (**VI, 10**) and a variety of sea-squirts such as the large, white, amorphous-looking *Phallusia mammillata* (**VI, 11**) which likes sheltered areas where there is little wave action.

A number of sea-anemones have adapted to conditions on muddy bottoms by evolving worm-like bodies which enable them to dig most of the body down into burrows in the mud. When undisturbed the tentacles are extended above the surface in wait for unwary prey, but when touched by something larger the tentacles retract and the anemone may disappear completely. The drab-coloured daisy anemone (*Cereus pedunculatus*) (**VI, 6**) sometimes adopts this mode of life, often where there are beds of eel-grass, although it may also be found attached to stones and not buried. The beautiful anemone *Edwardsia callimorpha* (**V, 11**), with its long pink body buried in the mud and tentacles spread out over the mud surface, also tends to be found in the vicinity of eel-grass, as well as elsewhere. The snakelocks anemone (*Anemonia sulcata*) (**VII, 20**), although occurring in a variety of situations, will often be found whenever eel-grass is present, attached to the leaves with the adhesive base of its column. (It is described in greater detail in the next chapter.) Another burrowing species, *Sagartiogeton undata* (**V, 8**), extends its long transparent tentacles well above the mud surface when covered by water and undisturbed—but the only chance of seeing this animal is on an extremely low spring tide as it occurs only at the very lowest parts of the shore and below.

The muddy shore provides a habitat for a wide variety of worms. The small yellow- and brown-striped flatworm *Prosthecereus vittatus* (**VI, 27**) may be common but is usually hidden under stones. In similar situations where there is gravel mixed with mud the red ribbon worm (*Lineus ruber*) (**VI, 8**) may be found from the middle shore downwards. The bristle worms however have colonised the muddy shore in the greatest numbers and have evolved a great many forms to fill a wide range of ecological niches. The peacock worm (*Sabella pavonina*) (**VI,**

9) and the dark violet *Myxicola infundibulum* (**V, 12**) are suspension feeders using their spectacular cilia-bearing fans to filter the water, but when disturbed they can snap into their tube in the mud in an instant.

One scavenger of the muddy shore well known to anglers is the ragworm (*Nereis diversicolor*) (**VI, 23**). This worm has chitinous jaws on the end of a protrusible proboscis that can suddenly be pushed out in front of the head. The ragworm takes live prey as well as small pieces of dead plant and animal material and this armoury is no doubt useful in capturing moving animals. It is a matter of interest that the ragworm also seems to be able to feed on suspended particles. It secretes a mucus net, which is placed near the entrance to its burrow, and then by undulations of the body it creates a water current through the burrow so that small particles carried by the current are trapped in the net. When a sufficient quantity of material has been collected in this way the net and particles are eaten and the whole process is repeated.

The paddle worms like *Phyllodoce maculata* (**V, 14**) with their leaf-life parapodia can swim as well as crawl and usually occur among rocks in muddy sand. Another worm, *Amphitrite johnstoni* (**V, 15**), lives in twisted tubes in the mud and extends its very long thread-like tentacles over the surface to collect its particulate food. These tentacles are often the only sign of the worm that can be seen when a stone is turned over, and careful digging is required to remove the delicate body without damage.

Finally, mention must be made of the sea-mouse *Aphrodite aculeata* (**VI, 21**) a most unlikely-looking species related to the scale worms. It is really an inhabitant of the sublittoral zone, but is found on the beach after storms which wash it ashore. It has been given its name because of the matted furry appearance created by the many fine bristles on its back which shine with brilliant iridescent golds, greens and reds in the light. The sea-mouse is a voracious carnivore feeding largely on other small worms.

On a muddy shore where there is no fresh or brackish water influence, and especially where eel-grass (*Zoster marina*) is present, the sea-cucumber or cotton-spinner (*Holothuria forskali*) (**VI, 5**) inhabits the lowest part of the shore so that it may be found only on the lowest spring tides. When disturbed, this echinoderm will frequently bring its startling defensive action into play, expelling a mass of sticky tubules from the anal region. If its attacker were a crab or a small lobster, this would entangle it and render it helpless, allowing the sea-cucumber to crawl away to safety. Another desperate habit of the animal when stressed is to eject the digestive tract, gonads and the respiratory apparatus through the anus. These are then regenerated by the animal if left alone.

Mudflats offer very little opportunity for shore fishes but small pools sometimes occur, for instance where tidal scour around a rock has created a depression, and these may be inhabited by a gunnel (*Pholis gunnellus*) (**II, 31**), sometimes known as the butterfish but not to be confused with the bivalve with the same common name. The common blenny or shanny (*Blennius pholis*) (**II, 26**) may also take refuge in one of these puddles, if stranded on the shore by the receding tide. From time to time species such as the pogge or armed bullhead (*Agonus cataphractus*) (**VI, 13**), which live on muddy bottoms in shallow water offshore, may become stranded in shallow pools low on the shore, and conger eels (*Conger conger*) are reported to turn up occasionally.

In the shallow waters of estuaries there are several fishes which are able to tolerate conditions of fluctuating salinity. Some of these are species which move into the estuary from fresh water or the sea from time to time and a few are passing migrants. More details, and illustrations, will be found in Chapter Three. The flounder (*Platichthys flesus*) is a common fish in estuaries where it may extend well into freshwater areas upstream; it also occurs in the freshwater parts of the Baltic. Its diet ranges from freshwater worms and molluscs when upstream to estuarine and marine crustaceans and worms when it is further down. Mature flounders which have grown in estuaries and rivers migrate into deeper waters offshore to spawn early in the year. During this time they lose a lot of weight as they do not feed much, depleting the reserves built up during the previous summer and autumn. Some, at least, of the spent fish return to the estuaries after spawning.

The grey mullets, of which there are several species, are usually common in estuaries; in warm weather they are often to be seen in the shallows and sometimes, where there are only a few inches of water, their backs and dorsal fins may break the surface. These fish browse on algae; in marinas, when it is quiet and they are undisturbed, the sucking noise they make as they remove algal growth from pontoons near the surface is clearly audible. The grey mullets feed in estuaries but breed in the open sea as does the bass (*Dicentrarchus labrax*), a predatory species which comes into estuaries in the summer.

The herring (*Clupea harengus*) is a marine species which migrates into estuaries in shoals from time to time; it is well able to withstand reduced salinities. It extends far into the Baltic where the freshwater influence is considerable and its distribution overlaps with that of the pike. In fact the pike is known to feed on herring in places where they occur together.

Towards the latter part of the summer mackerel (*Scomber scombrus*) may also enter estuaries for a short while. At this time of the year their diet consists largely of small fish, and the mouths of estuaries are generally a good hunting ground for species like the sprat (*Sprattus sprattus*).

Fishes passing through estuaries on their migrations include the salmon (*Salmo salar*), the life history of which is described in Chapter Three, and the common eel (*Anguilla anguilla*) (p. 81), which spawns in the Sargasso sea in mid-Atlantic. The leaf-shaped leptocephalus larvae which hatch from the eggs are carried by ocean currents to the coasts of north-west Europe where they metamorphose into the young adult form or elver before entering the estuaries and passing up into the fresh water beyond. They may spend from seven to twelve years in fresh water before returning to the sea.

An animal that does not fit exactly into the category of estuarine fish, but which does deserve mention here, is the lamprey. It is not regarded as a true fish because it possesses many primitive features. There are three species which occur in north-west Europe but only two, the lampern (*Lampetra fluviatalis*) and the sea lamprey (*Petromyzon marinus*) (p. 81) are likely to be seen in estuaries. They both live in the sea and migrate into fresh water via estuaries to breed in late winter to early summer; after spawning they die. In the sea they live a parasitic existence as bloodsuckers of fish such as cod, mackerel and herring.

The mass assemblages of molluscs, worms and small crustaceans on mudflats provide a source of food for large numbers of birds and the open expanses of many of these shores, offering little or no cover for predators, make them ideal feeding grounds. Migratory geese such as the brent (*Branta bernicla*) (**VI, 3**) and greylag (*Anser anser*) (**VI, 16**), however, have a vegetarian diet. The greylag breeds in Scotland but overwinters in the Mediterranean. There are two races of the brent; the dark-breasted overwinters on North Sea coasts and nests in Arctic Russia and Siberia, and the pale-breasted overwinters mainly in Ireland and nests in Greenland. The major diet of the brent goose is eel-grass and there has been a significant decline in their numbers since the eel-grass epidemic of the 1930s.

The waders are probably the most numerous birds on the mudflat. The curlew, greenshank, redshank and knot, already mentioned in the section on sandy shores, search here also at low tide, as well as many others like the dunlin (*Calidris alpina*) (**VI, 22**) and the black-tailed godwit (*Limosa limosa*) (p. 89). Eating approximately their own body weight per day they are significant predators of both mud and sand fauna. At high tide these birds often retreat from the shore but as the tide recedes they appear and make as much as possible of the opportunity to feed while their feeding ground is exposed.

Cormorants (*Phalacrocorax carbo*) (**VI, 2**) are common birds of estuaries as well as in the open sea (see also illustration in Chapter Three, **XII, 16**). They spend much of their time diving for fish, and as an adaptation to this way of life their feathers are less oily than those of other birds. This gives them the advantage that fewer air bubbles are trapped, providing greater streamlining and less buoyancy, but it means that when not fishing the cormorant has to stand for some time with the wings held out for the feathers to dry. This is a common sight on piles or other structures in estuaries. The shag (*Phalacrocorax aristotelis*) (p. 86) closely resembles the cormorant but it is smaller and is usually only found in open coastal places and rarely in estuaries or inland.

FOULING AND ANTI-FOULING

As most of our yacht marinas and moorings lie in the shallow estuaries and harbours of our coasts it is pertinent to mention here an aspect of the life of the shore and shallow waters which is of particular importance to the yachtsman. The organisms which cause fouling are all members of the shore community or from the shallow sublittoral zone. They are bottom-living species which have a planktonic phase in their life history and it is the planktonic larval stages which allow the rapid spread of fouling to any clean, untreated surface which is suspended in the sea.

The problems associated with marine organisms settling on the submerged areas of boats' hulls are only too familiar to the yachtsman. The slightest irregularity on the smooth surface of a yacht's hull can exert a considerable influence on performance, and so fouling, even in the early stages of growth, appreciably reduces the speed of a sailing yacht or increases the fuel consumption of a motor boat. Most of the organisms associated with fouling are dealt with elsewhere in this book—they are all forms of marine life which are adapted to live in natural marine habitats but which find that the hulls of boats provide a very suitable alternative to their normal settling surface.

The most important fouling organisms are various species of filamentous algae like *Enteromorpha intestinalis*, sometimes known as 'grass' in this context; the hydrozoan *Tubularia indivisa* which grows in colonies known as 'roses'; barnacles like the stalked goose barnacles *Lepas anatifera* and, more commonly, the acorn barnacles (*Balanus balanoides*); various tube worms; polyzoans or 'sea-mats'; mussels (*Mytilus edulis*) and several tunicates or 'sea-squirts'. To wooden hulls, boring species like the ship worm and the gribble are also a serious threat, not only to efficiency but to the very structure of the boat. The ship worm (*Teredo navalis*) (p. 77) is a bivalve mollusc in which the body has become worm-like and the shell is reduced to a small, highly specialised boring instrument; it bores into the timber and then turns at a right angle and bores with the grain. The burrow rarely breaks the surface and is lined with a shell-like substance. The gribble (*Limnoria lignorum*) (p. 77) is a small woodlouse-like crustacean which bores into timber to a depth of about 2·5 cm and then returns to the surface to select a fresh spot in which to bore again.

In the sailing ships of the past the hull below the waterline was sometimes protected with sheets of copper which prevented boring organisms making contact with the timber; the toxic nature of copper also discouraged settlement of plants and animals on the surface. Today anti-fouling paints have been developed which function by killing the larvae of fouling organisms prior to settlement, by the continuous release of toxic substances into the layer of water adjacent to the hull. Although anti-fouling compositions look like ordinary paint they should really be considered as a film acting as a toxic reservoir which must eventually become depleted and will fail when the biocide drops below the critical level necessary to kill the settling larvae. A number of biocides are used in different anti-fouling paints—these include cuprous oxide and metallic copper, organo-mercury, organo-arsenic, organo-tin compounds and organic compounds such as D.D.T. and T.M.T. Compounds of mercury, arsenic and D.D.T. are declining in popularity because of their toxicity to man and for environmental reasons.

Modern yacht anti-fouling can be divided into two basic categories. The first type is known as soluble matrix or soft film anti-fouling which contains soluble poisons and water-sensitive resins which are partly soluble in water. They work by releasing poison as the surface of the anti-fouling becomes porous and gradually wears away, exposing the lower layers. The second basic type is contact leaching or racing anti-fouling which provides a tough, smooth surface necessary for high-performance yachts. They are made from relatively water-resistant media containing high levels of biocide, and although the biocide is released into the surrounding water no erosion of the surface occurs.

PLANTS AND ANIMALS OF THE SEASHORE

Plant Kingdom

Division: Algae
Class: Chlorophycae (Green algae)

Ulva lactuca Linnaeus: Sea-lettuce. Green, semi-transparent, membranous frond up to 50 cm but usually smaller. Variable shape with small stalk providing attachment to rocks or other algae. Upper, middle and lower shore, often in pools. Atlantic, English Channel, North Sea, Baltic and Mediterranean. Common. (II, 29)

Enteromorpha intestinalis (Linnaeus) Link: Grass kelp. Pale green, unbranched, inflated fronds constricted irregularly along length. Usually 10–20 cm. Occurs in estuaries and particularly on shores where fresh water is present. Usually on upper shore in pools. Atlantic, English Channel, North Sea, Baltic and Mediterranean. Common. (II, 15)

Class: Phaeophycae (Brown algae)

Pelvitia canaliculata (Linnaeus) Decaisne & Thuret: Channelled wrack. Dark brown thallus, no midrib or bladders; swollen fruiting bodies on tips. Conspicuous groove on one side of frond. Up to 15 cm. Upper shore, forming a distinct zone on rocks. Atlantic, English Channel and North Sea. Common. (II, 6)

Fucus vesiculosus Linnaeus: Bladder wrack. Brown dichotomously branched thallus with midrib. Bladders often arranged in pairs. Yellow or brown fruiting bodies at tips. Up to 100 cm. Middle shore, forming a distinct zone on rocks. Occurs in estuaries on piles. Atlantic, English Channel, North Sea and Baltic. Common. (II, 5)

Fucus serratus Linnaeus: Serrated or toothed wrack. Brown, dichotomously branched thallus with midrib but no bladders; edges serrated. Pointed fruiting bodies in autumn and winter. Up to 60 cm.

Forms a zone on rocks on lower middle shore. Occurs in estuaries, Atlantic, English Channel, North Sea and Baltic. Common. (II, 2)

Fucus spiralis Linnaeus: Spiral wrack. Frond may be up to 40 cm in length. Tough and leathery, with an obvious midrib; tends to be twisted near the tips. Rounded reproductive bodies at the ends of branches. No air bladders. Upper shore on rocks where there is protection from severe wave action. Atlantic, English Channel and North Sea. Common. (p. 61)

Ascophyllum nodosum (Linnaeus) Le Jolis: Knotted or egg wrack. Olive-brown or green thallus up to 150 cm. Irregularly branched with egg-like bladders at intervals along centre of the frond; no midrib. Yellow fruiting bodies in early summer. Middle shore on rocks; occurs in estuaries. Atlantic, English Channel and North Sea. Common. (II, 3)

Halidrys siliquosa (Linnaeus) Lyngbye: Pod weed. Yellowish-brown alternately branched thallus with bladders which resemble small peapods. Small fruiting bodies on upper branches. Up to 150 cm. From lower middle shore to sublittoral on rocks. Atlantic, English Channel and North Sea. Common.

Laminaria digitata (Hudson) Lamouroux: Oarweed. Brown thallus consists of root-like holdfast, smooth stipe with oval cross section and strap-like fronds. 100–200 cm. Extreme lower shore and sublittoral on rocks. Atlantic, English Channel, North Sea and Baltic. Common (II, 4). *L. hyperborea* (Gunnerus): Red kelp. Similar to *L. digitata* but generally larger with rough, round section stipe.

Alaria esculenta (Linnaeus) Greville: Dabberlocks. Frond up to 30 cm or longer. Distinctive midrib supporting a long, thin, easily torn leaf. Lateral bunches of flattened

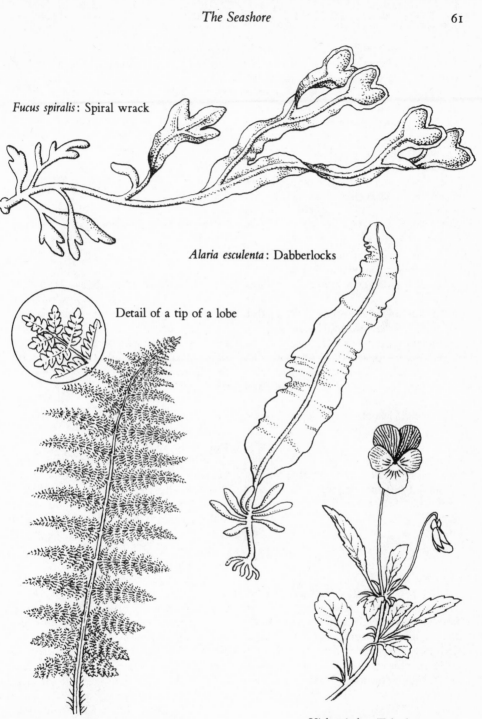

Fucus spiralis: Spiral wrack

Alaria esculenta: Dabberlocks

Detail of a tip of a lobe

Athyrium filix-femina: Lady fern

Viola tricolor: Tricolour pansy
or wild pansy

fruiting segments at base of stalk. Branching holdfast. On rocks on extreme lower shore in exposed places; may be the dominant algae in this position on the shore at very exposed sites. Northern Atlantic and North Sea. Common. (p. 61)

Class: Rhodophycae (Red algae)

Corallina officinalis Linnaeus: Coral weed. Pink calcareous thallus with jointed appearance; branches and branchlets opposite. Up to 10 cm. Middle shore on rocks often in pools. Atlantic, English Channel, North Sea, Baltic (occasionally) and Mediterranean. Common. (**II, 11**)

Lythophyllum incrustans Philippi. Pink calcareous encrusting thallus resembling icing sugar on shells and stones. Upper, middle and lower shores. Atlantic, English Channel, ·North Sea and Mediterranean. Common on exposed shores. (**II, 32**)

Rhodymenia palmata (Linnaeus) Greville: Dulse. Thallus a dark red flat frond branching dichotomously. Holdfast attached to laminarian stipes and rocks. Up to 30 cm. Middle and lower shore. Atlantic, English Channel and North Sea. Common. (**II, 7**)

Griffithsia flosculosa (Ellis) Batters. Thallus consists of delicate dichotomously branched threads. Pigmentation red with banded appearance. Fruiting bodies stalked dots on threads. Tangled holdfast. Lower shore. Atlantic, English Channel, North Sea, Baltic and Mediterranean. Common.

Division: Lichens

Cladonia foliacea (Huds.) Willd. Yellowish-green scale-like growth forming extensive mats. Scales deeply divided into segments which stand erect with tips bent when dry. Black bristles along margins. Sandy and gravel shores above high water mark. Atlantic, English Channel, North Sea, Mediterranean. Frequent. (**IV, 6**)

Xanthoria parietina (Linnaeus) T. A. Fries. Coarse leafy thallus, orange in colour. Encrusting on rocks in irregular patches up to 10 cm across. Forms a distinctly coloured belt on extreme upper shore. Atlantic, English Channel and North Sea. Abundant. (**I, 3**)

Verrucaria maura Acharius. Black encrusting thallus forming a distinct wide zone beneath *Xanthoria* and above the barnacles. May resemble an oil spill. Atlantic, English Channel and North Sea. Abundant. (**I, 4**)

Verrucaria mucosa Wahlenburg. Green-black encrusting thallus not more than 30 cm across. Appears gelatinous when wet. Common on rocky shores. (**I, 5**)

Division: Bryophyta (Mosses)

Polytrichum piliferum (Huds.). pale bluish-green leaves with greyish-white toothed points extending from tips; leaves crowded into a short terminal tuft at the end of the stem. Atlantic, English Channel, North Sea, Baltic, Mediterranean. Very common. (**IV, 2**)

Division: Pteridophyta (Ferns, etc.)

Athyrium filix-femina (Linnaeus) Roth: Lady fern. Short stout underground stem, densely clothed with brown pointed scales. Leaves form a crown at the apex of the rhizome; they die off in autumn. Found among shady rocks, also in damp woods and marshes. Throughout north-west Europe. Common in most districts. (p. 61)

Division: Gymnospermae

Picea abies (Linnaeus) Karst: Norwegian spruce. Tree up to 40 m high. Brown bark; twigs yellowish or reddish-brown; leaves 1–2 cm, many on each leaf cushion; cones 10–15 cm in length. Commonly planted tree for forestry. Native of northern and central Europe from Scandinavia and northern Russia to the Pyrenees and Italian alps. Common. (p. 63)

Pinus sylvestris Linnaeus: Scots pine. Tree up to 30 m high or occasionally 50 m. Old trees have a flat crown. Bark on trunk reddish-brown or orange, shed in scales. Twigs greenish-brown when young, becoming greyish-brown. 2 leaves on each short shoot. Cones 3–7 cm, dull brown with oblong scales. Scotland, Scandinavia and south to the mountains of Portugal and south Spain. Locally common. (p. 63)

Juniperus communis Linnaeus: Juniper. A shrub of which there are 2 subspecies. Subsp. *communis* is erect or spreading and prickly to the touch. Leaves are spreading and almost at right angles to the stem. This is the lowland form found locally often on chalk or limestone. It flowers May–June. Subsp. *nana* lies loosely along the ground; hardly prickly to the touch, the leaves wide and ascending. On rocks, moors and mountains. Locally common. (p. 63)

Division: Angiospermae (Flowering plants)

Cochlearia officinalis Linnaeus: Scurvy grass. White flowers May–August. Lower leaves heart-shaped with long stalks. Pods almost round. Stems 10–25 cm long. Above high water mark. Common. (I, 1)

Brassica oleracea Linnaeus: Wild cabbage. Yellow flowers on long spike May–August. Leaves very broad and covered with bluish-green bloom. Biennial. Height 30–60 cm. On rocky cliffs above high water mark. Not uncommon. (I, 9)

Cakile maritima Scop.: Sea rocket. Annual herb with prostrate or erect stem; long slender tap-root; 30–35 cm; leaves oblong lobes. Lilac or white flowers June–August. On drift line of sandy and shingle shores. Atlantic, English Channel, North Sea and western Baltic. Common. (III, 1)

Viola tricolor Linnaeus: Tricolour pansy or wild pansy. Usually annual. Flowers highly coloured, may vary from violet, mauve and yellow on 1 flower to 3 shades of mauve or mauve and white on the same clump; petals longer than sepals; flowering April–September. Locally common on waste ground and islands in British Isles and in Europe from Scandinavia to central France. (p. 61.) Subsp. *curtisii* (E. Forst.) Syme: Seaside pansy. Perennial, with yellow flowers. Found near the sea on

Diagram showing how conifers are able to establish a foothold on a rocky Baltic island.

Picea abies: Norwegian spruce
Pinus sylvestris: Scots pine
Detail of twin needles
Detail of needles
Detail of needles
Juniperus communis: Juniper

sandy dunes. West coast of Britain, Ireland, and shores of English Channel, North Sea and Baltic. Common.

Silene maritima (Withering) A. & D. Love: Sea campion. White flowers with broad petals June–August. Plant lying on ground but tending to rise at end. Bracts leaf-like. Above high water mark. The woody rootstocks may aid in stabilising shingles. Atlantic, English Channel and North Sea. Common. (**I, 7**)

Honkenya peploides (Linnaeus) Ehrh.: Sea-sandwort. Succulent herb with leafy flowering and non-flowering shoots 5–25 cm long. Fleshy, oval leaves. Flowering May–September. On sandy and stony shores; tolerant to short periods of immersion in seawater. Atlantic, English Channel and North Sea. Common. (**IV, 3**)

Spergularia marina (Linnaeus) Griseb: Sea-spurrey. Small pink flowers June–August. Spreading or prostrate branches; leaves flat on top and rounded beneath. Bracts often leafy; calyx longer than petals. A few of the seeds winged. Above high water mark. Common.

Spergularia media (Linnaeus) C. Presl.: Perennial sea-spurrey. Plant similar to *S. marina* but larger, and petals are longer than calyx. Pale pink flowers June–September. Usually found on muddy saltmarshes. Widespread. Common. (**V, 2**)

Spergularia rupicola Lebel ex Le Jolis: Rock sea-spurrey. Short-leaved plant flowering June–September. On sea cliffs and among rocks. Widespread. Common. (**I, 2**)

Sedum anglicum Huds.: English stonecrop. Prostrate, evergreen annual often tinged with red; numerous slender creeping and rooting stems for forming mats; vertical branches. Flowering June–August. On sand dunes, shingle and rocks. Atlantic, English Channel, North Sea, Baltic and Mediterranean. (**IV, 5**)

Eryngium maritimum Linnaeus: Sea-holly. Branched perennial 30–60 cm high. Very obvious bluish tint. Leaves spiny and with a thick cartilaginous margin. Blue flowers July–August. On sandy and shingle shores. Not common outside nature reserves. Atlantic, English Channel, North Sea, Baltic and Mediterranean. (**IV, 7**)

Crithmum maritimum Linnaeus: Samphire or rock samphire. Yellow flowers with spicy fragrance June–September. Fleshy leaves. Up to 30 cm with roots up to 100 cm tucked into rock crevices. Above high water mark. Atlantic and English Channel. Common. (**I, 8**)

Aster tripolium Linnaeus: Sea-aster. Stem up to 45 cm high with fleshy leaves. Flowering July–September. In saltmarshes. Atlantic, English Channel, North Sea, Baltic and Mediterranean coasts. Common. (**V, 4**)

Tripleurospermum maritimum (Linnaeus) Koch: Scentless chamomile. 2 subsp.: *maritimum* has short leaf segments while *inodorum* has long leaf segments. Subsp. *maritimum* flowers July–September. Coasts of Britain, western Europe and Baltic. Locally common.

Senecio jacobea Linnaeus: Ragwort. Biennial or perennial herb with erect flowering stems; 30–150 cm. Leaves smooth and lobed. Yellow flowers July–October. On sand dunes. Atlantic, English Channel, North Sea and Baltic. Very common. (**IV, 8**)

Limonium vulgare Mill.: Sea-lavender. Violet-blue flowers July–September. Large pinnately veined leaves often coloured red or yellow by rust fungus. Crowded spikelets. Calyx consists of lobes with teeth between them. Above higher water mark. Channel and North Sea. Common. (**V, 5**)

Limonium binervosum (G.E. Sm.) C.E. Salmon: Rock sea-lavender. Leaves elongate or slightly rounded. Flowering July–September. Widespread. Common. (**I, 6**)

Armeria maritima (Mill.) Willd: Thrift or sea-pink. Narrow leaves and pink flowers March–September. On saltmarshes, rocks and cliffs; also in mountainous areas.

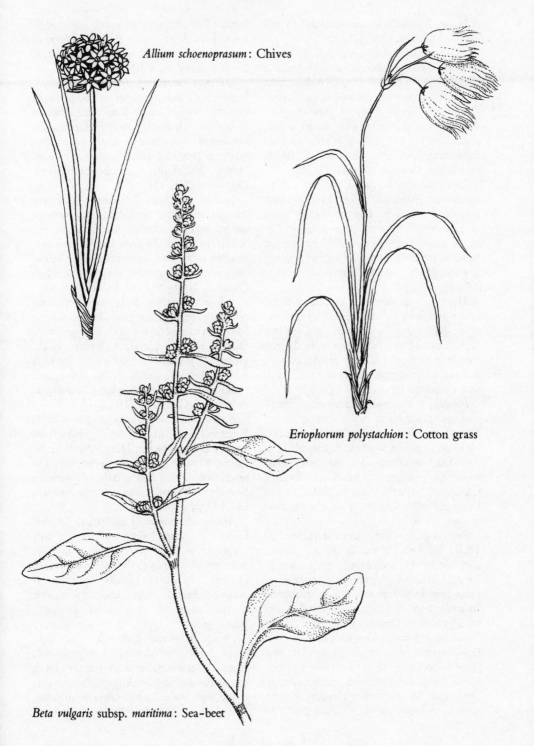

Allium schoenoprasum: Chives

Eriophorum polystachion: Cotton grass

Beta vulgaris subsp. *maritima*: Sea-beet

Atlantic, English Channel, North Sea and Baltic. Very common. (II, 1)

Glaux maritima Linnaeus: Sea milkwort or black saltwort. Perennial with trailing stem 10–30 cm long. Forms spreading mats on grassy parts of saltmarshes. Small pink flowers 0·5 cm in diameter June–August. Will occupy damp crevices under cliffs; once established will tolerate extreme conditions. Atlantic, English Channel, North Sea. Common locally. (V, 1)

Calystegia soldanella Linnaeus: Sea-bindweed. Perennial; lying loosely along ground; 10–60 cm. Funnel-shaped, pink or pale purple flowers June–August. On sandy and shingle shores. Atlantic (except western Ireland), English Channel, North Sea and Mediterranean. Common locally. (IV, 9)

Myosotis ramosissima Rochel: Early forget-me-not. Slender annual; 2–25 cm; lower leaves ovate and forming a rosette, upper leaves oblong; all leaves hairy. Bright blue flowers April–June. In dry areas with shallow soil. English Channel and North Sea. Common locally. (IV, 4)

Plantago maritima Linnaeus: Sea-plantain. Long linear fleshy leaves; brown flowers with pale yellow stamens. Flowering June–August. In saltmarshes, in short turf near the sea and also near mountain streams. Atlantic, English Channel, North Sea, Baltic and Mediterranean. Common on coasts, rare inland. (V, 6)

Beta vulgaris subsp. *maritima* (Linnaeus) Thell: Sea-beet. Stems lie on the ground and rise at the end; thick, green, glossy leaves; fruiting sepals are cork-like. Flowering July–September. On seashores. Atlantic, English Channel, North Sea and Mediterranean. Common. (p. 65)

Halimione portulacoides (Linnaeus) Sellen: Sea-purslane. Perennial; dense shrubby plant up to 80 cm or occasionally 150 cm. Stems lying along ground and tending to rise at end. Short creeping rhizome. Leaves opposite, flowering July–September. In saltmarshes fringing channels and pools flooded at high tides. Atlantic, English Channel, North Sea and Mediterranean. Locally abundant. (VI, 15)

Salicornia europaea Linnaeus: Glasswort. Much-branched stems up to 30 cm high; branches arise at a wide angle. Bright green fading to yellow or occasionally scarlet. Flowering August–September. In muddy tidal marshes. Atlantic, English Channel, North Sea, Baltic and Mediterranean. Common. (VI, 18)

Suaeda maritima (Linnaeus) Dumort: Annual sea-blite. Spreading, prostrate or sometimes erect annual; 30 cm, bluish and reddish in colour. Flowers July–September. In saltmarshes and seashores usually below high water spring tides. Atlantic, English Channel, North Sea and Baltic. (III, 3)

Salsola kali Linnaeus: Saltwort. Prostrate or rarely erect prickly annual up to 60 cm. Much-branched stems with pale green or reddish stripes; succulent leaves; thick, green sepals. Solitary flowers July–September. Sandy shores above high water mark. Atlantic, English Channel and North Sea. Common. (III, 4)

Euphorbia paralias Linnaeus: Sea-spurge. Perennial; 20–40 cm high with short woody stock; oblong, blunt, thick, fleshy leaves. Flowering August–September. On sandy shores and dunes. Atlantic, western English Channel and Mediterranean. Local. (IV, 1)

Allium schoenoprasum Linnaeus: Chives. Grows in tufts; cylindrical leaves; pale purple or pink flowers in June and July. Stamens $^1/_2$ as long as the spreading petals. On rocky pastures, usually limestone or granite. British Isles and north-west Europe south to north-west Portugal. Local. (p. 65)

Juncus maritimus Lamarck: Sea-rush. Erect, densely tufted, very tough perennial. Wiry and pointed. Stem 30–100 cm. Dark green. Flowering July–August. Saltmarshes above high water mark; often dominating large areas. Atlantic, English Channel,

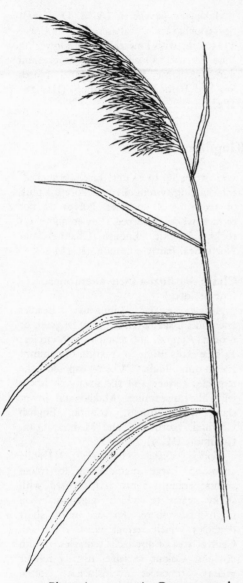

Phragmites communis: Common reed

North Sea, Baltic and Mediterranean. Very common. (**V, 3**)

Zostera marina Linnaeus: Eel-grass or common grass wrack. Broad, dark green, grass-like leaves with at least 3 veins; up to 100 cm long and 0·4–0·8 cm wide. Inconspicuous flowers June–September;

ribbed seeds. From extreme low water to approximately 4 m, on mud, fine gravel and sand. Rarely found in estuaries. Atlantic, English Channel, North Sea, Baltic and Mediterranean. Abundant locally. (**VI, 7**)

Eriophorum polystachion Linnaeus: Cotton grass. Leaves triangular and acutely angled, 3–6 mm wide. Overall height 20–60 cm. Found in wet boggy areas all over British Isles and north-west Europe. Common locally but has disappeared in some areas where drainage has destroyed the habitat. (p. 65)

Spartina × townsendii H. & J. Groves: Rice grass or Townsend's cord-grass. Stems 60–100 cm, leaves up to 0·8 cm wide. Flowering July–November. Arose as a fertile hybrid through crossing *S. alterniflora* and *S. maritima*. Has invaded many mudflats and estuaries and has been planted extensively to reclaim tidal areas as its matted roots and fast-spreading spear-like shoots quickly bind mud and build up the level. Common. (**VI, 14**)

Phragmites communis Trinius (= *P. australis* (Cav.) Streud.): Common reed. Stems 2–3 m; leaves 10–20 mm wide, tapering to a long slender point. Densely branched inflorescence. Freshwater reed which may grow at the inland extent of saltmarshes where conditions become fresh, and in swamps and shallow water inland. Atlantic, English Channel, North Sea, Baltic and Mediterranean. Very common. (p. 67)

Phleum arenarium Linnaeus: Sand cat's-tail. Annual grass 3–15 cm with short smooth leaves. Flowering May–July. On sand dunes, Atlantic, English Channel, North Sea, western Baltic and Mediterranean. Locally common. (**III, 5**)

Polypogon monspeliensis (Linnaeus) Desf.: Annual beard grass. Up to 60 cm with long leaves. Flowers July. Found in damp places near coasts and saltings. English Channel and Mediterranean. Not uncommon. (**V, 7**)

Ammophila arenaria (Linnaeus) Link:

Marram grass. Stout, erect perennial; 60–120 cm; wiry, tufted, in-rolled leaves and whitish flower spikes. Long rhizomes. Flowers July. On coastal sands where it is a major cause of sand dune formation. Atlantic, English Channel, North Sea and Baltic. Very common. (III, 2)

Agropyron junceiforme (A. & D. Love): Sea-couchgrass. Perennial grass, 25–50 cm, with long rhizomes. Flowering June–August. Occurs on young sand dunes. Atlantic, English Channel, North Sea and Baltic. Very common. (III, 18, 18a)

Animal Kingdom

Phylum; Porifera (Sponges)

Halichondria panicea (Pallas): Crumb-of-bread sponge. Encrusting sponge with smooth surface raised into volcano-like oscular openings. Colour variable: usually white, orange, yellow, green or brown. Under stones from middle shore to sublittoral. Atlantic, English Channel, North Sea, Baltic and Mediterranean. Common. (II, 28)

Hymeniacidon sanguinea (Grant). Encrusting sponge with very variable shape and texture, up to 50 cm across. Surface rough and covered with many oscular openings and grooves with no particular pattern. Coloured orange, red or scarlet. Under rocks and stones among mud from middle shore to low water. Atlantic, English Channel, North Sea and western Baltic. Common. (VI, 10)

Cliona celata Grant: Boring sponge. Smooth branching sponge usually only recognisable as holes in dead (or live) shells of *Venus* and *Ostrea* or rocks into which it has bored. The surface of the object may be riddled with these holes which are about 0·2 cm in diameter and through which a small portion of the yellow, green or blue sponge may be visible. Lower shore and below. Atlantic, English Channel, North Sea, western Baltic and Mediterranean. Common. (IV, 17)

Phylum: Coelenterata (Sea-firs, Sea-anemones, Corals, Jellyfish)
Class: Hydrozoa (Sea-firs)

Tubularia indivisa Linnaeus. Hydroid with stems up to 15 cm; rarely branched. Interlocking system of roots. Polyps a pink or red ring of tentacles. Polyps do not retract when disturbed. Lower shore on rocks. Atlantic, English Channel and North Sea. Fairly common. (I, 18)

Class: Anthozoa (Sea-anemones, Corals, etc.)

Actinia equina Linnaeus: Beadlet anemone. Usually red but may be green or brown. Approx. 200 tentacles in 6 circles; 24 blue spots inside top margin of column. 20–30 mm high. When exposed the tentacles retract and the anemone has a jelly-like appearance. Middle and lower shore under stones. Atlantic, English Channel, North Sea and Mediterranean. Common. (II, 9)

Tealia felina (Linnaeus): Dahlia anemone. Large anemone up to 10 cm across; column squat and covered with sticky warts to which sand and shell fragments adhere. 80 or more short tentacles which retract when disturbed. Central area of disc lacks tentacles, mouth obvious. Colour variable, usually red or green; tentacles translucent, often multicoloured. Lower shore to sublittoral. Atlantic, north to west Scotland, western English Channel and Mediterranean. Common. (II, 27)

Calliactis parasitica (Couch): 'Parasitic' anemone. Tall brown column up to 8 cm. Many fine yellow tentacles. Often found on empty shell of the whelk *Buccinum undatum* inhabited by the hermit crab

Eupagurus bernhardus. N.B. This species is not a parasite: the hermit crab benefits from the protection afforded by the anemone, which in turn benefits from the crab's activities. Extreme low shore. Atlantic, north to western Ireland, English Channel and Mediterranean. Not common. (**II, 12**)

Peachia hastata Gosse: Burrowing anemone. Worm-like anemone up to 10 cm long when fully extended. No adhesive base; column pink with bulb below a constriction near the bottom; 12 fine vertical lines. About 12 translucent tentacles patterned in shades of pale brown; characteristic arrow-like markings near their bases. Lower shore to 30 m; burrowing usually about 8 cm down in clean sand. Atlantic, English Channel, North Sea and Mediterranean. Not common. (**III, 31**)

Sagartia troglodytes (Price). Column up to 4 cm with wavy suckers and adhesive base. 100–200 tentacles up to 0·75 cm long. 2 varieties occur: *ornata* (**III, 27**), usually found on clean sand; disc up to 1 cm in diameter; column dark green; *decorata*, usually on muddy substratum; disc up to 3·5 cm in diameter; column yellow-grey. Lower shore and below. Atlantic, English Channel and North Sea. Common.

Cereus pedunculatus (Pennant): Daisy anemone. Elongated anemone with slender stalk up to 10 cm long and reaching 3 cm in diameter. Adhesive basal disc. Sometimes trumpet-shaped when fully extended. 300–700 short limp tentacles. Variable colour—pink, brown, cream or violet—with grey suckers near top of stalk either distributed irregularly or in rows. To these may be attached shell fragments, small stones or pieces of algae. Lower shore, buried in mud with tentacles exposed. Sometimes in estuaries. Atlantic, north to Scotland, English Channel and Mediterranean. Common. (**VI, 6**)

Sagartiogeton undata (Müller). Column a tall pillar up to 6 cm long with firmly adhesive base up to 4 cm in diameter; yellow with vertical brown stripes. Column able to contract to an almost flat mound. Oval disc brown with 100 fine transparent tentacles. Lower shore to sublittoral, buried in sandy mud. Atlantic, English Channel, North Sea, western Baltic and Mediterranean. (**V, 8**)

Edwardsia callimorpha (Gosse). Anemone with worm-like translucent pink body up to 10 cm high; no adhesive basal disc. 16 delicate tentacles marked with fine dots. Diameter across outspread tentacles 3·5 cm. Burrowing in muddy sand on lower shore and in shallow sublittoral. Atlantic, English Channel, North Sea and Mediterranean. Not uncommon. (**V, 11**)

Phylum: Platyhelminthes (Flatworms)

Prosthecereus vittatus (Montagu). Leaf-shaped flatworm with wavy edges up to 3 cm. Blunt head with 2 tentacles; pointed tail. Creamy white with longitudinal brown stripes. Lower shore and sublittoral under rocks and stones in mud. Atlantic, western English Channel and North Sea. Common locally. (**VI, 27**)

Phylum: Annelida (Segmented worms)
Class: Polychaeta

Aphrodite aculeata Linnaeus: Sea-mouse. Oval-shaped worm up to 10 cm long. 40 segments. Iridescent green and gold hairs on flanks; felt of brown hairs masks scales on dorsal surface. Lower shore and below. Atlantic, English Channel, North Sea, western Baltic and Mediterranean. Common, particularly after storms. (**VI, 21**)

Lepidonotus squamatus Linnaeus: Scale worm. Brownish yellow, 2 cm long. 26 body segments overlaid by 12 pairs of overlapping scales on dorsal surface, scales not rounded. Under stones on middle and lower shore. Atlantic, English Channel, North Sea, western Baltic. Common.

Nereis diversicolor O. F. Müller: Ragworm. Body up to 10 cm long with 90–120 segments possessing bristles. Very small antennae. Colour variable but usually yellowish-brown with green along sides. Distinct red line down back is the dorsal blood vessel. Middle shore to shallow sublittoral in mud and muddy sand. Atlantic, English Channel, North Sea, western Baltic and Mediterranean. Very common especially in estuaries. (**VI, 23**)

Nereis virens M. Sars: King ragworm. 20–40 cm long and thick as a finger. 100–175 segments with bristles. Jaws with 6–10 teeth each. Green with iridescent sheen above and iridescent pink beneath. Lower shore and below, burrowing in sand. Occurs in estuaries. Atlantic north of the English Channel, North Sea and western Baltic. Common locally. (**III, 14**)

Phyllodoce maculata (Linnaeus): Paddle worm. Greenish-yellow, up to 10 cm long. About 250 segments bearing brown-coloured paddles; 3 or 4 brown spots on each segment. Heart-shaped head with 4 antennae; eversible proboscis with no jaws. Lower shore in muddy sand under pebbles. Atlantic, English Channel, North Sea and western Baltic. Not common. (**V, 14**)

Nephtys hombergi Audouin and Milne Edwards: Cat-worm. Up to 10 cm long, occasionally 20 cm, with 90–200 segments. Pink or bluish-white with iridescent sheen; wide dark line down the back. Dense short bristles; 4 short antennae; long tail thread. Lower shore and below in sand and muddy gravel. Atlantic, English Channel, North Sea, western Baltic and Mediterranean. Common, locally abundant. (**III, 16**)

Chaetopterus variopedatus (Renier): Parchment worm. 22–25 cm long, wide head and 2 long antennae; large mouth. 1st 9 segments bear bristles; 12th segment bears wing-like processes; waist region has 3 segments on which are carried leaf-like paddles; posterior segments bear long parapodia. Very fragile U-tubes in which this worm lives are up to 40 cm long and 1·5 cm wide; they are parchment-like with adhering sand grains. Lower shore and below, in sand and muddy sand. Atlantic, English Channel, North Sea and Mediterranean. (**IV, 12**)

Lanice conchilega (Pallas): Sand mason. Worm up to 15 cm or occasionally longer. Simple tube of hardened mucus covered with grains of sand and shell fragments; branched like a tree at anterior end. Body possesses 150–300 segments; 17 segments possess bristles; dull red gills and pale pink tentacles at anterior end; thoracic region swollen, abdomen thin and fragile. Middle shore and below, in sand. Atlantic, English Channel, North Sea and Mediterranean. Common, abundant locally. (**III, 17**)

Amphitrite johnstoni Malmgren. Fat soft worm up to 20 cm, tapering towards the tail. Crown of very long orange or pink thread tentacles arise from the head and 3 pairs of dark red gills just behind the head. 90–100 segments, 24 bearing bristles. Buff-yellow or brown. Tube of mucus, buried in mud or under stones. Lower shore. Atlantic, English Channel, North Sea and Mediterranean. Common. (**V, 15**)

Myxicola infundibulum (Renier). Up to 12 cm long, flattened with 100–150 segments; yellow or orange. A dark brown or violet fan of feathered tentacles connected by a fine membrane may be quickly withdrawn when disturbed. Tube transparent and jelly-like, up to 15 cm long and 3 cm across; buried completely in mud. Lower shore and sublittoral in mud and muddy gravel. Atlantic, English Channel, North Sea and Mediterranean. Common locally. (**V, 12**)

Sabella pavonina Savigny: Peacock worm. 10–25 cm long with 100–600 small segments; rounded above and flattened below. Fan consists of 2 semi-circular clusters of 8–45 filaments arising from 2 distinct lobes giving the impression of a single crown; brown, red or violet with dark bands. Fan quickly withdraws when disturbed or when exposed at low tide.

Smooth round tube of fine mud up to 45 cm long and 0·75 cm in diameter, sticking 7–10 cm out of the ground. Lower shore on muddy flats. Atlantic, English Channel, North Sea and Mediterranean. Common. (**VI, 9**)

Arenicola marina (Linnaeus): Common lugworm. Up to 20 cm long. Soft slimy cylindrical body. 6 swollen anterior segments then 13 segments each bearing a pair of red gills. Middle shore and below, burrowing in sand. Occurs in estuaries. Atlantic, English Channel, North Sea and western Baltic. Common. (**III, 13**)

Spirorbis borealis Daudin. Small, off-white calcareous tubes coiled clockwise; up to 3 mm across. When placed in water, worm inside extends green tentacles. Middle and lower shore on fucoids, laminarians or rocks. May occur in huge numbers. Atlantic, English Channel, North Sea and Mediterranean. Very common. (**II, 2**)

Phylum: Priapuloidea

Priapulis caudatus Lamarck. Plump, cylindrical, gherkin-shaped body up to 8 cm long. Club-like eversible proboscis with spines at anterior end and tassel-like tail. Lower shore and below, in sand and mud. Atlantic, North Sea and western Baltic. Rare. (**IV, 10**)

Phylum: Nemertina (Ribbon worms)

Lineus ruber O. F. Müller: Red ribbon worm. Up to 16 cm long with flattened head bearing 3–4 eyes in a row on each side. Body flattened and tapering towards tail; distinct 'neck'. Reddish-brown or pink; ventral surface lighter than dorsal. Middle shore to sublittoral, under stones in muddy gravel; sometimes several together. Atlantic, English Channel, North Sea, Baltic and Mediterranean. Common. (**VI, 8**)

Phylum: Mollusca (Molluscs)
Class: Polyplacophora

Lepidochitona cinereus (Linnaeus): Grey chiton or coat-of-mail shell. Dull grey chiton up to 2 cm long. Fleshy edge around shell plates coloured red, brown or green. 16–19 pairs of gills. Width approximately $^3/_4$ of the length. Difficult to distinguish from other species but this is generally the commonest. Upper, middle and lower shores under stones. Atlantic, English Channel, North Sea, south-western Baltic and Mediterranean. Common. (**I, 12**)

Class: Gastropoda (Snails and slugs)

Patella vulgata Linnaeus: Common limpet. Tall, blunt limpet up to 7 cm but usually 3–4 cm. Greenish-blue or grey shell with white-yellow interior. White or brown scar inside shell apex. Transparent tentacles fringe the mantle when the animal is submerged and moving. Upper and middle shore on rocks. Atlantic, English Channel and North Sea. Very common. (**II, 19**)

Buccinum undatum Linnaeus: Common whelk or buckie. Pale brown shell up to 8 cm long. 7–8 whorls with deep sutures. Opening yellow with smooth margin. Short siphonal canal. Seen under water the extended siphon is long. Egg capsules, commonly found on shore, consist of yellowish sponge-like masses. Lower shore and sublittoral, on sand and mud. Atlantic, English Channel, North Sea and western Baltic. Common. (**II, 12**)

Ocenebra erinacea (Linnaeus): Sting winkle or oyster drill. Shell up to 5 cm high with 8–10 whorls. Well-marked sutures; longitudinal and spiral ridges. Siphonal canal open in juveniles but closed in adults. Outer lip of opening toothed. Yellowish-white with brown markings. Lower shore and below; migrates onshore to spawn. Atlantic, English Channel and North Sea. Common, particularly on oyster beds where it is a pest. (**VI, 26**)

Nucella lapillus (Linnaeus): Common dogwhelk. Heavy shell up to 4 cm with about 5 whorls. Flattened spiral ribs. Short siphonal canal. Yellow or white sometimes with dark bands. Middle shore on rocks. Atlantic, English Channel and North Sea. Very common, often in large numbers associated with barnacles or mussels. (**II, 16**)

Nassarius reticulatus (Linnaeus): Netted dogwhelk. Conical shell up to 3 cm high with about 7 whorls. Network of longitudinal and diagonal ribs produces pattern of small squares. Aperture oval with teeth on outer lip. Short siphonal canal. Brown, grey or white. Lower shore and sublittoral. Under stones especially where sand and mud present. Atlantic, English Channel, North Sea, western Baltic and Mediterranean. Common.

Gibbula cineraria (Linnaeus): Grey topshell. Shell up to 1·25 cm high, and of similar width. Up to 7 whorls; small umbilicus. Grey shell with red bands which may be faint. Among stones and often on seaweeds on lower shore and below. Atlantic, English Channel and North Sea. Very common. (p. 73)

Calliostoma zizyphinum (Linnaeus): Painted topshell. Cone-shaped shell with straight sides; no umbilicus. 2·5–3 cm in height. Yellow and pink with red stripes. Lower shore and sublittoral, among rocks. Atlantic, English Channel, North Sea and Mediterranean. Common. (**II, 18**)

Littorina littorea (Linnaeus): Common or edible periwinkle. Shell up to 2·5 cm high. Sharp conical shape. Grey or green with concentric darker bands. Middle shore, on rocks and in estuaries. Atlantic, English Channel, North Sea, western Baltic and Mediterranean. Very common. (**II, 21**)

Littorina littoralis (Linnaeus): Flat periwinkle. Shell up to 1 cm high with flat or slightly rounded spire, very finely sculptured surface. Large aperture. Various distinct colour forms occur: yellow, orange, green, red, purple, brown or streaked. Middle and lower shore, on seaweeds. Atlantic, English Channel, North Sea and western Baltic. Very common. (**II, 20**)

Littorina neritoides (Linnaeus): Small periwinkle. Shell about 0·5 cm high and sharply pointed. Surface of shell smooth. Bluish-black. On upper shore in cracks and crevices especially where the beach is exposed to wave action. Atlantic, English Channel, North Sea and Mediterranean. Very common. (p. 73)

Littorina saxatalis (Olivi) (= *L. rudis*): Rough periwinkle. Shell about 0·8 cm high. 6–9 whorls with deep sutures in between. Shell rough to the touch. Usually reddish, yellow or greyish in colour. In cracks and crevices and on stones on upper and upper middle shore. Atlantic, English Channel, North Sea and western Baltic. Very common. (p. 73)

Apporhais pespelicani (Linnaeus): Pelican's foot shell. Shell up to 3 cm high. Last whorl heavily ridged and drawn out into characteristic lip. Greyish-yellow. Lower shore into deeper water, burrowing in sand and mud or on the substratum surface. Atlantic, English Channel and North Sea. Not uncommon. (**III, 12**)

Turritella communis (Risso): Tower shell. Long tapering shell up to 5 cm high. 19 whorls with spiral ridges. Relatively small, rounded opening. Reddish-brown, yellow or white. Extreme lower shore and below on mud and muddy gravel. Live animals partially buried but dead shells are often found on the surface. Atlantic, English Channel, North Sea and Mediterranean. Common, locally abundant. (**VI, 25**)

Crepidula fornicata (Linnaeus): Slipper limpet. Oval-shaped shell up to 4 cm with coiling evident at one end. Projecting plate on inside of shell. Grey or whitish with brown markings. Often found in chains with largest individuals at the bottom. These are females and the smaller individuals at the top are males. The males change sex as they grow. Lower shore and shallow water on oyster beds where they are a pest competing for space. Occurs in estuaries. Atlantic, English Channel and North Sea where it has been introduced

Gibbula cineraria:
Grey topshell

Littorina neritoides:
Small periwinkle

Littorina saxatalis:
Rough periwinkle

Cochlicella acuta: Sand-hill or pointed snail

Dorsal   Lateral

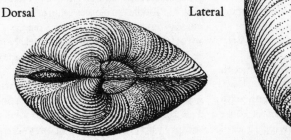

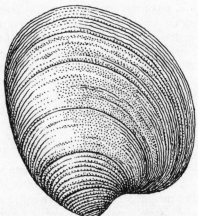

Mercenaria mercenaria: American hard-shell clam or quahog

from America. Common, abundant locally. (**III, 11**)

Hydrobia ulvae (Pennant): Laver spire shell. Elongate shell up to 0·6 cm high with 6 whorls ending in a point. Outer lip of the opening straight-edged where it joins the spire at an acute angle. Brown or brownish-yellow. Middle shore in estuaries and in saltmarshes. May be so abundant as to give the mud surface a granular appearance. Atlantic, English Channel, North Sea and Baltic. Very common. (**VI, 24**)

Archidoris pseudoargus (Rapp): Sea-lemon. Elliptically-shaped body up to 7 cm long. Warty yellow skin with brown or green blotches. 2 tentacles at the anterior end and the anus on the dorsal surface towards the posterior end is surrounded by

a ring of 9 plumed retractile gills. Lower shore, in summer only. Atlantic, English Channel and North Sea. Very common. (**I, 15**)

Cochlicella acuta (Miller): Sand-hill snail or pointed snail. Shell a narrow pointed spire 15–20 mm high; thin-walled; rounded whorls; minute umbilicus. Usually whitish with brown bands across the whorls but colour variable. Near the coast; Atlantic, English Channel and Mediterranean; usually on sand dunes but also inland in dry areas of Ireland. Common.

Class: Bivalvia

Mytilus edulis Linnaeus: Common mussel. Black or dark brown shell occasionally with darker radial bands;

rarely more than 10 cm long. Hinge of shell does not possess teeth but 3–12 crenulations occur beneath the beaks. Beaks at end of shell. Shell margin smooth. Inside of shell white with dark border. Anterior adductor muscle scar small, posterior large. Middle to lower shore on rocks, attached by the bysuss threads; in estuaries on piles. Atlantic, English Channel, North Sea, Baltic and Mediterranean. Common, particularly on exposed rocky shores. (II, 23)

Cerastoderma edule (Linnaeus): Common cockle. Oval shell with similar valves up to 5 cm long but usually less. Margin of shell notched; 22–26 coarse radiating ribs. White, brownish-yellow or grey. Lower shore, burrowing in sand or mud. Occurs in estuaries. Atlantic, English Channel, North Sea and Mediterranean. Very common. (III, 10)

Dosinia lupinus (Linnaeus): Smooth artemis. Rounded shell up to 3·5 cm long. Smooth concentric ridges. Yellowish with white interior. Long siphons. Lower shore and into deeper water, burrowing in sand and shell gravel. Atlantic, English Channel, North Sea and Mediterranean. Common. (III, 7)

Venus verrucosa Linnaeus: Warty venus shell. Rounded solid shell up to 6 cm long. Strong concentric ridges which end in warty tubercles at posterior end. Inside edge of shell notched. Yellow, white, grey or brown. Lower shore and into deeper water, burrowing in sand and gravel. Atlantic, English Channel and Mediterranean. Fairly common. (III, 24)

Venus striatula (da Costa): Striped venus. Solid shell up to 4·5 cm long. More or less triangular in outline with rounded anterior margin and slightly drawn out posteriorly. Dirty white, cream or pale yellow usually with 3 red-brown rays radiating from the beaks. In clean and muddy sand from the lower shore to 55 m. Atlantic, English Channel, North Sea and Mediterranean. Common. (IV, 17)

Venus casina Linnaeus. Rounded solid shell up to 5 cm long. Prominent concentric ridges interspersed with smaller ridges. Inside edge of shell notched. Whitish, sometimes with red-brown rays outside and white inside. Extreme lower shore and into deeper water, burrowing in sand and gravel. Atlantic, English Channel, North Sea and Mediterranean. Common.

Venerupis decussata (Linnaeus): Butterfish or cross-cut carpet shell. Solid shell up to 7 cm long. Broadly oval in outline but with a sharp curve behind the beaks. White, yellow or light brown outside; inside polished white with an orange tint. Sculpturing on shell consists of concentric grooves crossed and broken by radiating ridges. Middle to lower shore, in mud, muddy gravel and sand; occurs in estuaries. Atlantic, English Channel, North Sea (occasionally) and Mediterranean. Common. (VI, 30)

Mercenaria mercenaria Linnaeus: American hard-shell clam or quahog. Solid shell up to 12 cm long. Broadly oval in outline. Dirty white, grey or grey-brown outside; inside white, sometimes with violet in the region of the muscle scars. 3 cardinal teeth near the beaks on each valve. Concentric ridges cover the shell but these may be worn in older specimens. Upper middle shore to below low water, in mud. Introduced from the U.S.A. into various estuaries in Britain, France, the Netherlands and Belgium. Abundant locally. (p. 73)

Mya arenaria Linnaeus: Gaper shell. Shell up to 15 cm long with left valve less convex than right. Oval outline with posterior drawn out to the right and gaping. Large projecting spoon-like process on left valve to which ligament is attached when alive. Brown or dirty white outside, white inside shell. 2 adductor scars. Lower shore and below, burrowing in mud or sand; occurs in estuaries. Atlantic, English Channel, North Sea and Baltic. Common. (VI, 31)

Lutraria lutraria (Linnaeus): Common otter shell. Oval shell up to 12 cm long; gaping at each end. Teeth form a V-shaped projection on the left valve which fit in a corresponding socket on the right valve. 2 muscle scars; internal and external ligament; long siphons. Periostracum brown. Lower shore and below, burrowing in mud, sand or gravel. Atlantic, English Channel, North Sea and Mediterranean. Common. (**VI, 12**)

Ostrea edulis Linnaeus: Common European or flat oyster. Generally rounded shell but shape variable. Up to 100 cm. Left valve cupped and usually attached to other shell or rock, right valve flat and usually uppermost. No teeth on hinge; ligament internal; single adductor muscle scar. Greyish-brown or white. Lower shore and below; sometimes in dense commercial beds. Atlantic, English Channel, North Sea and Mediterranean. Common locally. (**VI, 29**)

Anomia ephippium Linnaeus: Common saddle oyster. Circular flat shell up to 6 cm long. Lower shell possesses an aperture through which a calcified bysuss extends, attaching the shell to surface such as a rock or other shell. Upper shell often assumes the shape of the surface of attachment. Distinct mother-of-pearl lustre inside shell. Middle shore to sublittoral. Atlantic, English Channel, North Sea and Mediterranean. Common, locally very abundant. (**II, 24**)

Scrobicularia plana (da Costa): Peppery furrow shell. Solid but light shell up to 6·35 cm long. Broadly oval in outline and much flattened. Grey, light brown or light yellow outside; inside white. Sculpture of irregular concentric lines and ridges. 2 cardinal teeth on right valve, 1 on the left. Long siphons which leave star-shaped pattern on mud surface. Upper middle shore to low water mark; occurs in estuaries. Atlantic, English Channel, North Sea, Baltic and Mediterranean. Common. (**VI, 20**)

Macoma balthica (Linnaeus): Baltic tellin. Solid shell up to 2·5 cm long. Anterior end rounded and slightly inflated, posterior end drawn out a little. White, yellow, pink or purple, often in concentric bands outside; inside white, purple or the same as outside. Shell surface smooth. 2 small cardinal teeth on each valve. Upper shore to low water, in mud, muddy sand and muddy gravel particularly in estuaries and areas of reduced salinity. Atlantic, English Channel, North Sea and Baltic. Common. (**VI, 19**)

Tellina tenuis da Costa: Thin tellin. Flattened shell with thin valves which are almost the same; up to 2 cm long. Bright glossy colours: yellow, red, orange, pink or white; often banded; colour inside the same as outside. Middle shore to shallow water, burrowing in sand. Atlantic, English Channel, North Sea, Baltic and Mediterranean. Very common. (**III, 20**)

Donax vittatus (da Costa): Banded wedge shell. Elongated wedge-shaped shell with equal valves up to 3·75 cm long. Margin coarsely notched. Glossy periostracum. Brown, purple or yellow. Middle shore to medium depths, burrowing in clean sand. Atlantic, English Channel, North Sea and Mediterranean. Common, locally abundant. (**III, 19**)

Gari depressa (Pennant): Large sunset shell. Oval shell up to 6 cm long with almost equal valves which gape at posterior end. Brown periostracum which may be worn away in places. Reddish-brown rays on outer area of shell. Lower shore to deeper water burrowing in sand. Atlantic, English Channel, North Sea and Mediterranean. Common locally. (**IV, 15**)

Ensis ensis (Linnaeus): Razor shell. Long narrow shell up to 12 cm long; slightly curved. Valves similar. Beaks very reduced. Whitish with brownish-green periostracum. Lower shore and shallow water, burrowing in sand. May be collected at low spring tides on moonlit nights. Atlantic, English Channel, North Sea and Mediterranean. Common. (**III, 25**)

Ensis siliqua (Linnaeus): Pod razor shell. Long, narrow, straight shell up to 20 cm long; does not taper towards the posterior end. Whitish in colour and covered with fine lines vertically and horizontally; glossy yellowish-green periostracum. Found burrowing in sand on the extreme lower shore and down to water depths of 35 m. Atlantic, English Channel, North Sea and Mediterranean. Common. (p. 77)

Spisula solida (Linnaeus): Thick trough shell. Triangular-shaped shell up to 4·5 cm long. Valves similar and gaping slightly at posterior end. Short siphons with a horny sheath. Yellowish-white with smooth concentric grooves. Extreme lower shore to deeper water, burrowing in sand and gravel. Atlantic, English Channel and North Sea. Common. (**III, 9**)

Pholas dactylus Linnaeus: Common piddock. Brittle shell up to 15 cm long; it has an inflated appearance and the 2 valves are similar. Edge of shell is smooth except at anterior end where there are crenulations. Front end of shell gapes; has a sculpturing of radiating and concentric ribs. Pale yellow periostracum; shell whitish-grey outside, white inside. Bores into soft rock, clay, wood, firm sand or peat on the lower shore and to a water depth of a few metres. Atlantic as far north as south-west Britain and Ireland, English Channel and Mediterranean. Common. (p. 77)

Zirfaea crispata (Linnaeus): Oval piddock. Shell up to 9 cm long, shorter and more compact than that of *Pholas dactylus*. Edge of shell is smooth except at anterior end. In life the valves hardly meet; they gape widely at the front and back end. White in colour. Found boring into peat, clay, shale, sandstone and other soft rock but rarely into wood. Atlantic to as far south as Biscay, English Channel and the North Sea. Common. (p. 77)

Teredo navalis Linnaeus: Ship worm. Worm-like bivalve whose shell has been reduced to a drilling organ for boring into wood. The mantle secretes a hard chalky tube up to 20 cm long which lines the hole. White in colour. Bores into pilings and boats and is frequently found in driftwood. Atlantic, English Channel, North Sea, western Baltic and Mediterranean. Common. (p. 77)

Phylum: Arthropoda (Arthropods)
Class: Crustacea
Subclass: Cirripedia (Barnacles)

Balanus balanoides (Linnaeus): Acorn barnacle. Dirty white barnacle up to 1·5 cm across. 6 plates fuse to form shell. Encrusting on rocks, often in large numbers. In the opening, sutures join mid-line obliquely. Upper shore and below, occurs in estuaries. Atlantic, English Channel, North Sea and western Baltic. Very common. (**II, 22.**) *Chthalamus stellatus* (Poli): Star barnacle. Very similar to *B. balanoides* but sutures in the opening join midline at right angles. A more southerly species.

Subclass: Malacostraca

Neomysis integer (Leach). Mysid shrimp up to 1·7 cm. Thin carapace, long abdomen with 6 segments, the last of which is not quite covered by the carapace. Large eyes on stalks. Tail fan. In pools and lagoons in estuaries. Atlantic, English Channel, North Sea and Baltic. Common. (**V, 10**)

Ligia oceanica (Linnaeus): Sea-slater. Flat, greenish-grey woodlouse-like crustacean up to 2·5 cm long. Stout outer antennae about $^2/_3$ body length. Two uropods each dividing into 2 long points. Immediately above intertidal zone in cracks and crevices. Active at night. Atlantic, English Channel, North Sea and Mediterranean. Common. (**I, 11**)

Limnoria lignorum (Rathke): Gribble. Body 0·35 cm long with parallel sides and rounded ends; 2 pairs of antennae of equal length; thorax slightly longer and wider than the abdomen; last segment of 6 abdominal segments slightly hollowed above. Rolls into a ball when exposed.

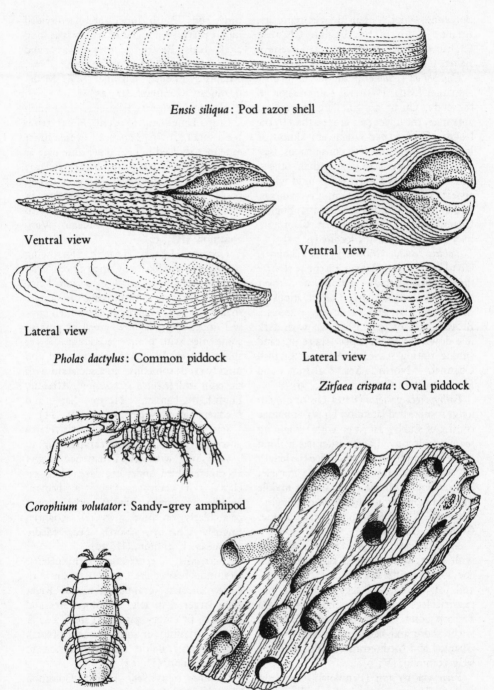

Ensis siliqua: Pod razor shell

Ventral view

Ventral view

Lateral view

Lateral view

Pholas dactylus: Common piddock

Zirfaea crispata: Oval piddock

Corophium volutator: Sandy-grey amphipod

Limnoria lignorum: Gribble

Teredo navalis: Ship worm

Bores into wood; often in large numbers. Atlantic north of the English Channel, English Channel, North Sea and western Baltic. Common. (p. 77)

Corophium volutator (Pallas). Sandy-grey amphipod with no lateral compression of the body. Up to 0·8 cm long. 2 pairs of antennae; the upper pair are less than $^1/_2$ the length of the lower which are almost as long as the body and very conspicuous. Last pair of walking legs longer than others. Lives in U-shaped burrows in mud which appear as holes 0·2–0·3 cm in diameter. Middle shore, in estuaries. Atlantic, English Channel, North Sea and Baltic. Common, may be very abundant locally (p. 77).

Talitris saltator (Montagu): Sandhopper. Up to 1·6 cm long. Upper antennae shorter than long-jointed section of lower antennae. Rough, toothed articulations to small joints at end of lower antennae. Brown, grey or greenish-fawn with dark line down the back. Upper shore on sand among rotting seaweed. Atlantic, English Channel, North Sea, Baltic and Mediterranean. Common. (**III, 29**)

Bathyporeia pelagica (Bate). Up to 0·6 cm long, compressed laterally. Upper antennae consist of stubby 1st joint with remaining segments droopy. Lower antennae as long as the body in the male and twice the length of the upper antennae in the female. Translucent with red eyes. Lower middle shore, burrowing in sand. Atlantic, English Channel, North Sea and western Baltic. Common. (**IV, 16**)

Squilla desmaresti Risso. Up to 12 cm long with short, shield-shaped carapace which does not cover the last 4 thoracic segments. 2nd pair of thoracic legs developed into raptorial legs with 4 conspicuous spines and a sharp point on the last joint. Extreme lower shore and below. Atlantic, English Channel and Mediterranean. Not particularly common. (**V, 9**)

Palaemon serratus (Pennant): Common prawn. Transparent prawn with purple markings. Up to 6 cm. Outer antennae $1^1/_2$ times body length. Inner antennae divided into 3 parts, the longest of which is as long as the body. Rostrum curves upwards and ends in 2 points. In pools on lower shore. Atlantic, English Channel and Mediterranean. Common. (**II, 25**)

Crangon vulgaris Fabricus: Common shrimp. Length up to 5 cm. No rostrum but small tooth between eyes. Inner antennae divided into 2; outer antennae as long as the body. 1st pair of legs carry large nippers. Lower shore and shallow water in pools where there is sand. Occurs in estuaries. Atlantic, English Channel, North Sea, Baltic and Mediterranean. Very common. (**III, 30**)

Upogebia deltaura (Leach): Burrowing prawn. Prawn-like body up to 10 cm long. Small head with small eyes and hairy rostrum. Pincers with moving part longer than fixed part. Hairy legs. Abdomen curved under body. Yellow-green or pinkish sometimes with orange markings. Lower shore and below, burrowing in sand and muddy sand; sometimes in association with the coin shell (*Lepton squamosum*). Atlantic, English Channel, North Sea and Mediterranean. Not common. (**IV, 11**)

Porcellana platycheles (Pennant): Hairy porcelain crab. Rounded carapace up to 1·2 cm long with broad, flattened claws; abdomen folded under carapace. Hairs on claws and carapace. Grey or brown. Middle to lower shore under stones particularly on muddy gravel. Atlantic, English Channel, North Sea, Mediterranean. Common. (**II, 30**)

Eupagurus bernhardus (Linnaeus): Common hermit crab. Up to 10 cm with large, unequal, granulated claws. Right claw larger than left. Abdomen soft and coiled to fit empty gastropod shell. Lower shore. Atlantic, English Channel, North Sea, western Baltic and Mediterranean. Very common. (**II, 12**)

Galathea squamifera Leach: Common squat lobster. Body up to 4·5 cm long. 1st pair of walking legs bear pincers and are

$1^1/_2$ times the body length. 4 pairs of lateral points on the rostrum, the hindmost being the smallest. Green-brown with red markings. Lower shore, down to 80 m under rocks and in cracks. Atlantic, English Channel, North Sea and Mediterranean. Common. (**I, 10**)

Carcinus maenas (Linnaeus): Common shore crab. Dark green, up to 4 cm long. 3 blunt projections between eyes and 5 sharp projections on side of carapace. Powerful claws, last joint of back legs flat and pointed. Middle and lower shore. Occurs in estuaries. Atlantic, English Channel, North Sea, Baltic and Mediterranean. Very common. (**II, 34**)

Macropipus puber Linnaeus: Fiddler or velvet swimming crab. Red-brown crab covered with fine brown hair. Up to 8 cm long. Red eyes; 8–10 projections between eyes, 5 projections on side of carapace. Last joint of legs flattened into swimming paddles. Lower shore. Atlantic, English Channel, North Sea. Common.

Corystes cassivelaunus (Pennant): Masked crab. Length up to 4 cm. Shell longer than it is broad. Long hairy antennae held closely together. Nippers in male twice as long as body; less than $^1/_2$ this length in the female. Lower shore and shallow water, buried in sand. Atlantic, English Channel and North Sea. Not uncommon. (**III, 28**)

Potamobius astacus (Linnaeus): European crayfish. Female up to 12 cm long, male up to 16 cm. Body and claws broad; antennae shorter than body length. In the male the 1st 2 pairs of tail legs form tubes. Colour is greenish-brown, blue or somewhat red. In lakes and rivers of north-west Europe and introduced into Britain. May extend to the sea in Baltic areas. Probably more common than is generally thought. (p. 81)

Nymphon gracile Leach: Sea-spider. Extremely thin body up to 1 cm long with 4 slender walking legs up to 2·5 cm. Abdomen very small. Pincer-like feeding appendages. Middle and lower shore. Atlantic, English Channel and North Sea. Com-

mon. (Unrelated to true spiders.) (p. 81)

Pycnogonum littorale (Stom): Sea-spider. Relatively heavy body with knobbly appearance. Thick, clearly segmented legs with strong claws. Yellow-brown in colour. Lower shore under stones. Atlantic, English Channel and North Sea. Not uncommon. (**II, 33**)

Class: Insecta

Philonicus albiceps (Meigen): Greater dune robber-fly. Grey-brown, bristly fly with a long horny proboscis. Often found crouching on bare sand and making short darts into the air to capture prey insects in flight. Widespread; common. (p. 81)

Phylum: Polyzoa (= Bryozoa or Ectoprocta)

Membranipora membranacea (Linnaeus): Sea-mat. Mat-like encrusting colony on seaweeds, particularly laminarians, stones and shells. Variable size and shape but may be up to 20 cm across. Generally white or grey in colour. Zooids rectangular with a blunt bristle on each side. Middle shore to below low water, on all types of substratum. Atlantic, English Channel, North Sea and Mediterranean. Common. (**V, 16**)

Flustra foliacea (Linnaeus): Hornwrack. Greyish-yellow plant-like colony up to 20 cm in height. Horny texture but flexible. Branched and often split at ends. Under the microscope the zooids appear rectangular and the zooecium possesses 2 pairs of bristles at one end. From lower shore to sublittoral, on rocks and stones. Atlantic, English Channel, North Sea and Mediterranean. Common. (**I, 17**)

Phylum: Echinodermata (Echinoderms)
Class: Asteroidea (Starfish)

Asterias rubens Linnaeus: Common starfish. Total diameter up to 50 cm but more usually 5–10 cm. 5 plump, tapering arms with blunt spines irregularly spaced over upper surface; sometimes 4 or 6 arms. Tube feet beneath the arms bear suckers. Yellow-brown or red-brown with yellow and occasionally violet markings. Lower shore to sublittoral, among rocks and stones. Often a pest on oyster beds. Atlantic, English Channel, North Sea and western Baltic. Very common. (II, 14)

Luidia ciliaris (Philippi). Large starfish up to 50 cm or more in diameter. 7 flattened arms which are often unequal in length. The body is depressed and slightly domed on top; disc is small. Tube feet do not possess suckers but end in knobs. Orange or red above, white below. Extreme lower shore and below, on mud and sand; often buried. Atlantic, English Channel, North Sea and Mediterranean. Fairly common. (V, 13)

Astropecten irregularis (Pennant): Burrowing starfish. Up to 12 cm across with 5 arms which have spiny appearance due to 2 layers of spine-bearing marginal plates. The tube feet do not possess suckers. Orange, brown or purple above, white underneath. Extreme lower shore and below, burrowing in sand. Atlantic, English Channel, North Sea, western Baltic and Mediterranean. Not uncommon. (III, 26)

Asterina gibbosa (Pennant): Cushion-star or starlet. Up to 5 cm but usually about 2 cm in diameter. Body plump and pentagonal, arms rounded at ends and turned up when out of water. Stiff, rough texture over upper surface. Green-yellow with cream tints. Lower shore and sublittoral, on rocks and under stones. Atlantic, English Channel and Mediterranean. Common. (II, 8)

Class: Ophiuroidea (Brittle stars)

Ophiura texturata Lamarck. Brittle star with central disc up to 3 cm across with 2 obvious plates near the base of each arm. Tapering arms up to 4 times the diameter of the disc. Brownish-orange. Lower shore and into deeper water, burrowing in sand.

Atlantic, English Channel, North Sea, western Baltic and Mediterranean. Common. (IV, 14)

Class: Echinoidea (Sea-urchins)

Psammechinus miliaris (Gmelin): Green sea-urchin. Up to 4 cm in diameter with short strong spines; green with purple tips. Test green when cleaned. Often disguised with pieces of weed or shell. Lower middle shore to sublittoral, on rocks, under stones and sometimes associated with oyster beds. Atlantic, English Channel, North Sea and western Baltic. Common locally. (I, 16)

Echinocardium cordatum (Pennant): Sea-potato. Heart-shaped urchin up to 6 cm long. 5 rows of tube feet; anterior row lying in a deep furrow. Dense fur-like spines shorter on top with occasional longer spines which are curved. Yellowish-brown when alive but white when dead and without spines. Lower shore and into deeper water. Atlantic, English Channel, North Sea and Mediterranean. Common. (III, 15)

Class: Holothuroidea (Sea-cucumbers)

Holothuria forskali Delle Chiaje: Sea-cucumber or cotton-spinner. Cucumber-shaped body up to 20 cm long. Skin soft and thick with coarse texture and warty appearance. Upper surface rounded, dark brown and bearing 2 rows of tube feet with suckers. Lower surface lighter in colour and bearing 3 rows of tube feet with strong suckers. Around the mouth are 20 yellow modified tube feet giving the appearance of tentacles; these are retractile. Glands near the anus eject white sticky threads when the animal is disturbed. Lower shore and below, on mud and muddy sand, particularly among *Zostera*. Atlantic and English Channel. Uncommon. (VI, 5)

Cucumaria normani Pace: Sea-cucumber. Up to 15 cm but usually less. Leathery and gherkin-shaped with dirty-brown-coloured skin. Locomotory tube feet arranged in

Plate III

FORMS OF WILDLIFE ASSOCIATED WITH SANDY SHORES

1. *Cakile maritima* : Sea-rocket
2. *Ammophila arenaria* : Marram grass
3. *Suaeda maritima* : Annual sea-blite
4. *Salsola kali* : Saltwort
5. *Phleum arenarium* : Sand cat's-tail grass
6. Cuttle bone
7. *Dosinia lupinus* : Smooth artemis
8. Dried egg-case of the thornback ray (*Raja clavata*)
9. *Spisula solida* : Thick trough shell
10. *Cerastoderma edule* : Common cockle
11. *Crepidula fornicata* : Slipper limpet
12. *Apforhais pespelicani* : Pelican's foot shell
13. *Arenicola marina* : Common lugworm
14. *Nereis virens* : King ragworm
15. *Echinocardium cordatum* : Sea-potato
16. *Nephtys hombergi* : Cat-worm
17. *Lanice conchilega* : Sand mason

18

22

19

20

21

23

24

25

26

27

28

29

30

31

Plate III

FORMS OF WILDLIFE ASSOCIATED WITH SANDY SHORES

18. and 18 (a) Marram grass and sea-couchgrass (*Agropyron junceiforme*) in process of consolidating a sand dune
19. *Donax vittatus*: Banded wedge shell
20. *Tellina tenuis*: Thin tellin shell
21. *Charadrius hiaticula*: Common ringed plover
22. *Tadorna tadorna*: Common shelduck
23. *Haematopus ostralegus*: Oystercatcher
24. *Venus verrucosa*: Warty venus shell
25. *Ensis ensis*: Razor shell
26. *Astropecten irregularis*: Burrowing starfish
27. *Sagartia troglodytes*: Sea-anemone
28. *Corystes cassivelaunus*: Masked crab
29. *Talitris saltator*: Sandhopper
30. *Crangon vulgaris*: Common shrimp
31. *Peachia hastata*: Burrowing anemone

Plate IV

PLANTS AND ANIMALS OF SANDY SHORES

1. *Euphorbia paralias*: Sea-spurge
2. *Polytrichum piliferum*: Hair moss
3. *Honkenya peploides*: Sea-sandwort
4. *Myosotis ramosissima*: Early forget-me-not
5. *Sedum anglicum*: English stonecrop
6. *Cladonia foliacea*: Dune lichen
7. *Eryngium maritimum*: Sea-holly
8. *Senecio jacobea*: Ragwort
9. *Calystegia soldanella*: Sea-bindweed

1

2

3

4

5

6

7

8

9

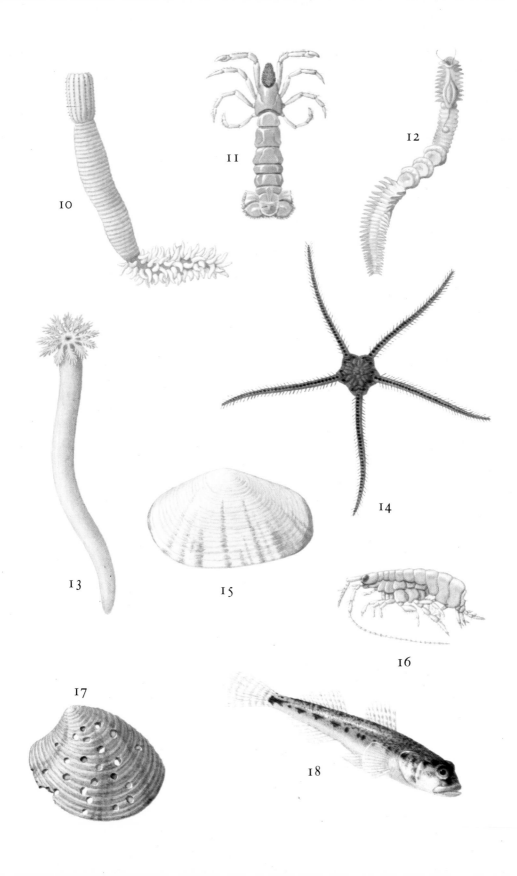

Plate IV

PLANTS AND ANIMALS OF SANDY SHORES

10. *Priapulus caudatus*
11. *Upogebia deltaura* : Burrowing prawn
12. *Chaetopterus variopedatus* : Parchment worm
13. *Leptosynapta inhaerens* : Worm-cucumber
14. *Ophiura texturata* : Brittle star
15. *Gari depressa* : Large sunset shell
16. *Bathyporeia pelagica* : Amphipod
17. *Venus striatula* : Striped venus shell, which has been bored into by *Cliona celata*, the boring sponge
18. *Gobius minutus* : Sand goby

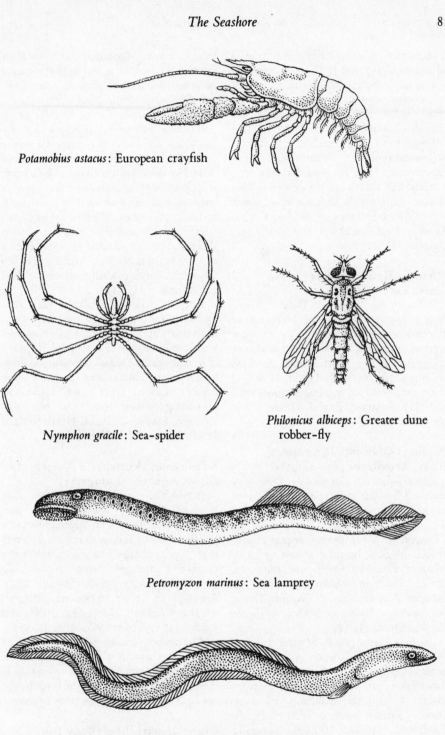

Potamobius astacus: European crayfish

Nymphon gracile: Sea-spider

Philonicus albiceps: Greater dune robber-fly

Petromyzon marinus: Sea lamprey

Anguilla anguilla: Common eel

double rows on lower surface but reduced on upper. Under water, 10 specialised tube feet around the mouth extend as feather projections. Lower shore to sublittoral, under rocks and stones. Atlantic and English Channel. Unusual. (**I, 13**)

Leptosynapta inhaerens (O. F. Müller): Worm-cucumber. Worm-like sea-cucumber up to 18 cm long. No tube feet. Anchor-like skeletal structures embedded in skin. 12 tentacles. Extreme lower shore and below, burrowing in sand and mud. Atlantic, English Channel and North Sea. Unusual. (**IV, 13**)

Phylum: Hemichordata
Class: Enteropneusta

Ballanoglossus clavigerus Delle Chiaje: Acorn worm. Worm-like body up to 30 cm divided into 3 regions. Short yellow anterior proboscis, behind which is a short colar and a long pale brown flattened abdomen. No tentacles. Lower shore and below, in U-shaped burrows in mud, sand and clay. Atlantic, English Channel and Mediterranean. Unusual. (**V, 17**)

Phylum: Chordata (Chordates)
Class: Ascidiacea (Sea-squirts)

Ciona intestinalis (Linnaeus): Tube sea-squirt. Jelly-like cylindrical body up to 12 cm high attached to rocks, seaweed piles and boat hulls at base. Grey, green or yellowish with red internal organs clearly visible. Inhalent opening at top of body, exhalent close by; both are retractile. Lower shore to sublittoral, occurs in estuaries. Atlantic, English Channel, North Sea, western Baltic and Mediterranean. Very common. (**I, 14**)

Ascidiella aspera (O. F. Müller). Solitary sea-squirt but may occur in unattached groups. Up to 10 cm long. Cylindrical or sometimes pear-shaped. Rough texture, slightly transparent; whitish, yellow or brown. Surface covered with unevenly distributed papillae. Inhalent opening terminal, exhalent opening $\frac{1}{3}$ of body length away. Openings on retractile siphons. Lower shore and below, in mud on stones, seaweeds or piles. Atlantic, English Channel, North Sea, western Baltic and Mediterranean. Common. (**VI, 28**)

Phallusia mammillata (Curier). Solitary sea-squirt up to 15 cm long. Generally cucumber-shaped with many conspicuous wart-like swellings and lumps. Thick tunic with texture of cartilage. Inhalent opening terminal with exhalent opening less than $\frac{1}{2}$ body length away. Slightly transparent; white or yellow-brown. Extreme lower shore and below, attached to stones buried in mud or muddy sand. Atlantic, English Channel and Mediterranean. Not uncommon. (**VI, 11**)

Botryllus schlosseri (Pallas): Star sea-squirt. Star-shaped groups of 6–20 individuals up to 2 mm each form colonies with common cloaca embedded in jelly-like translucent mass. Brown, yellow, green or red. Stars often in contrasting colour. Lower shore and sublittoral, encrusting stones, rocks and seaweeds. Atlantic, English Channel, North Sea and Mediterranean. Common. (**II, 10**)

Subphylum: Vertebrata (Vertebrates)
Class: Agnatha (Lampreys)

Petromyzon marinus Linnaeus: Sea lamprey. Up to 75 cm long, sometimes more. Eel-like body with 2 dorsal fins and a tail fin; smooth skin with no scales. Small eyes; 7 pairs of gills. The mouth is the most distinctive feature, being round and sucker-like with many small teeth. Migrates up rivers to spawn. Atlantic, English Channel, North Sea, Baltic and Mediterranean. Not uncommon. (p. 81)

The lampern (*Lampetra fluviatilis*) occurs in the same area. It is distinguished by the presence of a tooth plate at the upper edge of the disc with 2 widely separated teeth; the sea lamprey has numerous rings of teeth.

Class: Osteichthyes (Bony fish)

Sprattus sprattus (Linnaeus): Sprat. Up to

16 cm long. Herring-like fish with flattened body; belly has a sharply toothed keel; large scales which are easily lost from the body. Young sprat are known as whitebait. In coastal waters and estuaries. Atlantic, English Channel, North Sea, Baltic and Mediterranean. Very common.

Blennius pholis Linnaeus: Common blenny or shanny. Up to 15 cm. Head rises sharply above snout, no tentacles above eye as in other blennies but inconspicuous tentacles on cheeks. Yellow, olive or dark green with black markings. Lower shore, among rocks and stones. Atlantic, English Channel and North Sea. Commonest blenny. (**II, 26**)

Gobius minutus Pallas (= *Pomatoschistus minutus*): Sand goby. Up to 9 cm long. Slender fish with a long tail stalk. Sandy-brown with dark dots and 4 indistinct saddle-like markings. Males have a dark white-rimmed spot on the posterior edge of the 1st dorsal fin. On inshore sandy ground from the middle shore to 10 m below low water. Atlantic, English Channel, North Sea, western Baltic and Mediterranean. Very common. (**IV, 18**)

Pholis gunnellus (Linnaeus): Gunnel or butterfish. Up to 20 cm with rounded head and thick lips. Grey-brown with lighter markings and 9–13 dark spots surrounded by pale rings along base of long dorsal fin. Dark stripe from corner of mouth to eye. Lower shore to sublittoral, among rocks and weed, in pools and in estuaries. Atlantic north of the English Channel, English Channel, North Sea and western Baltic. Very common. (**II, 31**)

Agonus cataphractus (Linnaeus): Pogge or armed bullhead. Length up to 15 cm. Large triangular spiny head with short white barbels on gullet. Body tapers to a long tail stalk. Large pectoral fins, 2 dorsal fins; body encased in bony plates. Brownish-grey with 4 dark spots or stripes on dorsal surface; ventral surface whitish. Extreme lower shore and below, on mud and sand, sometimes buried. Atlantic north of the

English Channel, English Channel, North Sea and western Baltic. Not particularly common. (**VI, 13**)

Spinachia spinachia (Linnaeus): Fifteen-spined stickleback. Up to 16 cm with tapering head and finely tapering tail. Sides of body raised into a ridge of approx. 40 keeled plates. Dorsal fin preceded by about 15 spines. Olive-green above, silvery beneath; clear silver line runs forward below eyes with dark band above passing through eye to dark line on gill covers. Lower shore in pools, occurs in estuaries. Atlantic, English Channel, North Sea and Baltic. Common. (**II, 13**)

Conger conger (Linnaeus): Conger eel. Length up to 2 m but usually 30–60 cm. Grey above, pale below. Large mouth. Dorsal fin starts above pectorals and extends backwards along length of body. No pelvic fins; no scales; prominent gill slits. Occasionally found on muddy shores and below low water among rocks and in wrecks. Atlantic, English Channel, North Sea, western Baltic and Mediterranean. Common.

Anguilla anguilla (Linnaeus): Common eel. Up to 1 m long; males usually smaller. Very elongate body with continuous dorsal, tail and anal fins. Small gill opening in front of pectoral fin. No pelvic fins. Very small scales embedded in slimy skin. Can be considered a sea fish which spends much of its life in fresh water. Atlantic, English Channel, North Sea and Baltic. Common. (p. 81)

Esox lucius Linnaeus: Pike. Up to 1 m long. Unmistakable fish with extended snout and long jaws; single dorsal fin which is slightly forked. In freshwater rivers and lakes and low salinity areas of the Baltic. Common. (p. 84)

Stizostedion lucioperca (Linnaeus): Pike-perch or zander. Length 35–55 cm. Slender body; dorsal fin has 13–15 spines; dark elongated spots on dorsal fin form broken stripes. Mouth has several large fangs and many small teeth; in large fish the upper

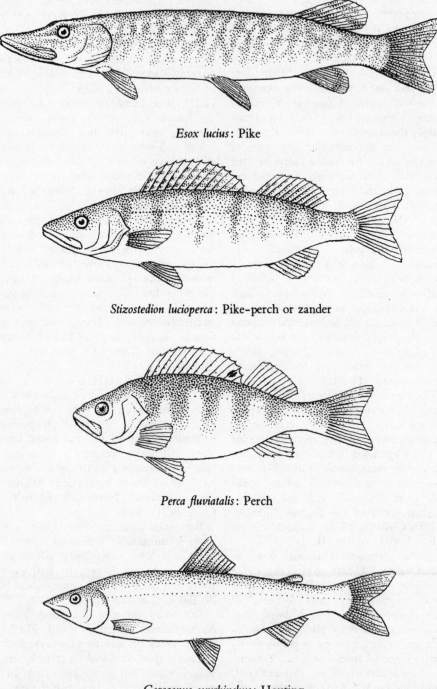

Esox lucius: Pike

Stizostedion lucioperca: Pike-perch or zander

Perca fluviatilis: Perch

Coregonus oxyrhinchus: Houting

jaw-bone extends back beyond the centre of the eye. The gill cover does not possess a spine as in the perch. Found in large lakes and the lower reaches of rivers; may extend to the sea in the Baltic. Found in many areas of north-west Europe and has been introduced to Britain. Common. (p. 84)

Perca fluviatilis Linnaeus: Perch. Length usually up to 25 cm but sometimes more. Hump-backed body; dorsal fin has 13–15 spines and a black spot at the posterior end. Strong spine on the gill cover. Greenish coloration which is darker in shallow-living specimens. In lakes, rivers and brackish waters such as the Baltic and in the Danish sounds. Found in most of north-west Europe and the British Isles. Common. (p. 84)

Coregonus oxyrhinchus (Linnaeus): Houting. Up to 50 cm long. Very variable whitefish; slender and silvery, resembling the herring apart from the adipose fin which is typical of the salmonids. Deeply forked tail fin and small mouth. 1st gill arch has 35–44 rakers. Northern Europe and Asia in lakes and rivers and in inshore Baltic waters. A migratory form is found on North Sea coasts from where it wanders up rivers. Also in Alpine and Swedish lakes. Not uncommon. (p. 84)

Class: Aves (Birds)

Podiceps cristatus Linnaeus: Great crested grebe. Length 45 cm. In the breeding season has a double black crest on the head and chestnut-coloured frills on the neck; these are lost in winter. Long white neck and pink bill. Sexes alike. Nest on or close to water. Mutual courtship display includes diving for waterweed and standing breast to breast on the water. Nests exclusively near fresh water but may be seen along seashores in sheltered bays particularly in winter. Atlantic, English Channel, North Sea, Baltic and Mediterranean. Increasingly common. (p. 86)

Podiceps grisegena Boddaert: Red-necked grebe. Length 43 cm. Similar to the great crested grebe in winter but with a shorter, darker neck and shorter black and yellow bill. In summer its red neck should make it unmistakable. Call: 'kell-kell' and a squeal in the breeding season. On marshy lakes and ponds and mainly on salt water in winter. Southern North Sea and Baltic. Common in Baltic; scarce but regular winter visitor to Britain. (p. 86)

Podiceps auritus Linnaeus: Slavonian or horned grebe. Length 34 cm. In summer unmistakable with golden ear tufts and chestnut-coloured neck. In winter tufts vanish and neck, breast and cheeks are white. Sexes alike. Call: a rippling trill. Nests inland beside lakes east of the Baltic; in winter found in estuaries and on sheltered coasts. Atlantic, English Channel, North Sea, Baltic and parts of western Mediterranean. Uncommon. (p. 86)

Ardea cinerea Linnaeus: Heron or grey heron. Length 91 cm. Upper parts grey; dark grey flight feathers; black crest; bushy breastplate and sharply pointed yellow bill. Characteristic flight with legs trailing and head drawn in. Sexes alike. Call: a harsh 'kaark'. Nest usually in established heronries in tall trees but also in reed beds or on cliffs. On rivers, lakes, estuaries and the seashore. Atlantic, English Channel, North Sea, southern Baltic and Mediterranean. Common. (p. 86)

Anas platyrhynchos Linnaeus: Mallard. Length 58 cm. Drake has glossy green head; white collar and purple-brown breast. Duck is brown. Both possess purple patch on the wing. Call: a loud quack by the drake; duck quieter. Nest on the ground; lined with down. Ducklings tended by duck alone. Surface feeder which reaches the bottom by 'up-ending' in shallow fresh and brackish waters. Atlantic, English Channel, North Sea, Baltic and Mediterranean. Very common. (p.86)

Phalacrocorax carbo (Linnaeus): Cormorant. Length 90 cm. Bluish-black with white chin. During the breeding season white patches are visible on the thighs.

Winter

Podiceps cristatus: Great crested grebe

Winter

Podiceps grisegena: Red-necked grebe

Winter

Podiceps auritus: Slavonian or horned grebe

Ardea cinerea: Heron or grey heron

Anas platyrhynchos: Mallard

Phalacrocorax aristotelis: Shag

Large beak curved downwards at the tip. Usually silent but may produce guttural groans when on the breeding ground. Nests colonially on rocky cliffs. Occurs in estuaries and on open coasts. Atlantic, English Channel, North Sea, western Baltic and Mediterranean. Common. (**VI, 2.**) The shag (*Phalacrocorax aristotelis*) (Linnaeus) is similar but smaller (length 62 cm). It has no white on the chin or thighs and there is a crest on the head in spring and summer. Plumage a glossy green-black. Atlantic, English Channel, North Sea and Mediterranean. Becoming increasingly common. (p. 86)

Aythya fuligula (Linnaeus): Tufted duck. Length 43 cm. Drake has a long droopy crest; glossy black plumage with contrasting white flanks. Duck is brown with smaller crest. White wing-bar apparent in both sexes in flight. Nests on the ground near waterside. A diving duck, found on the seashore in large numbers during winter. Atlantic, English Channel, North Sea, Baltic and Mediterranean. Common. (p. 88)

Bucephala clangula (Linnaeus): Golden-eye. Length 47 cm. Drake has a black and white body; black head with green sheen and a white circle in front of the eye. Duck is grey with brown head, white collar and wing patches. Quiet, but duck may make a grunting noise. Normally nests in tree holes in the Baltic area. Found on lakes and rivers but winters on the coast. Atlantic, English Channel, North Sea, Baltic and Mediterranean. Common in the northern areas. (p. 88)

Tadorna tadorna (Linnaeus): Common shelduck. Length 60 cm. Black, white and chestnut plumage; red bill with knob on base in the drake. Sexes alike. Nests in burrows vacated by rabbits. Call: 'ag-ag-ag'. A shore duck, feeding on crabs and mussels, etc. Atlantic, English Channel, North Sea, western Baltic (summer) and Mediterranean (winter). Common. (**III, 22**)

Anser anser (Linnaeus): Greylag goose. Length 80 cm. Barred grey-brown back; pale grey forewing; large orange bill. Sexes alike. Generally heavy appearance. Call: 'honk', similar to domestic goose of which it is the ancestor. Seen on the coast, particularly in the winter when it favours mudflats and saltmarshes. Breeding north Scotland, north and central Europe, wintering coasts of Britain, Holland, Portugal and Adriatic. Atlantic, English Channel, North Sea and northern Baltic in summer only. Common. (**VI, 16**)

Anser fabalis (Latham): Bean goose. Length 76 cm. Dark head and neck; long black and orange bill; yellow legs. Sexes alike. Call: a low honk. Nest a scrape in the ground with leaves, moss and down; in northern Europe. Found near the North Sea and Baltic in winter but takes to tundra and open northern woods in summer. Not widespread. (p. 88)

Branta bernicla (Linnaeus): Brent goose. Length 58 cm. Black head, neck and upper breast; small white patch on neck; dark grey back and wings; black tail with white above and below. 2 races: dark-breasted with slate-grey under parts, nesting in Russia and Siberia, and pale-breasted with white under parts, nesting in Spitzbergen and Greenland. Sexes alike. Call: a deep 'honk'. Rarely seen inland; roosts on the sea. Atlantic (pale-breasted race, largely confined to west of Ireland and north-east England), English Channel and North Sea. Common locally, where *Zostera* beds grow. (**VI, 3**)

Cygnus olor (Gmelin): Mute swan. Length 1·5 m. White plumage; orange bill with black base and a black knob which is smaller in the female. Long neck carried curved in an S-shape. Sexes similar. Juveniles or cygnets have a dirty-brown appearance. Nests near fresh water on the ground. In winter often found on seashores in sheltered bays. Atlantic, English Channel, North Sea and Baltic. Common. (p. 88)

Aythya fuligula: Tufted duck

Female

Bucephala clangula: Goldeneye

Anser fabalis: Bean goose

Lyrurus tetrix: Black grouse

Cygnus cygnus: Whooper swan

Cygnus columbianus: Bewick's swan

Cygnus olor: Mute swan

Charadrius alexandrinus: Kentish plover

Arenaria interpres: Turnstone

Tringa nebularia: Greenshank

Limosa limosa: Black-tailed godwit

Calidris canutus: Knot

Corvus corone cornix: Hooded or grey crow

Cygnus cygnus (Linnaeus): Whooper swan. Length 1·5 m. White plumage; long bill, black at tip and yellow at base. Sexes alike. Call: a noisy bugle-like whooping. Nests on islands, swamps and lakes in the north but in winter prefers salt water. Atlantic coasts of Ireland and Britain, North Sea and Baltic. Not uncommon. (p. 88)

Cygnus columbianus Yarrell: Bewick's swan. Length 1·2 m. Similar to whooper swan but smaller and possessing a shorter bill with a rounded yellow patch at the base. Sexes alike. Nests in summer in Arctic tundra but in winter found on lakes, larger rivers and sheltered sea bays. Irish coast, North Sea and southern Baltic. The least common of the swans. (p. 88)

Lyrurus tetrix (Linnaeus): Black grouse. Length: male 53 cm, female 38 cm. Male has lyre-shaped tail with white underfeathers; glossy blue-black plumage with white wing-bar. Female is smaller with chestnut and buff plumage and a forked tail. Nests on the ground—a scrape made by the female; males polygamous and take no part in nesting. On heath and moors and open woods. May be seen on some Baltic islands. Northern Europe and Britain. Fairly common. (p. 88)

Haematopus ostralegus (Linnaeus): Oystercatcher. Length 43 cm. Black and white with conspicuous long orange bill and pink legs. Sexes alike. Call: a noisy 'bleep'. Feeds on the shore, probing into mud and taking shellfish from rocks. May nest inland. Atlantic, English Channel, North Sea, Baltic (summer) and Mediterranean (winter). Very common. (**III, 23**)

Charadrius hiaticula (Linnaeus): Ringed plover. Length 18 cm. Black and white head with conspicuous black collar. Short bill; yellow legs; prominent wing-bar in flight. Sexes similar. Call: 'choo-ee'. Feed in groups on the shore. Atlantic, English Channel, North Sea and Baltic. Common. (**III, 21**)

Charadrius alexandrinus Linnaeus: Kentish plover. Length 15 cm. Upper parts pale brown; incomplete neck band; white wing-bars in flight; black legs. Call: a low pitched 'chu-uu-ee'. Winters in Africa, fairly common in summer on sandy beaches in France, Belgium, Holland, Denmark and in the Mediterranean. Rare spring passage migrant on southern part of the east coast of England. (p. 88)

Arenaria interpres (Linnaeus): Turnstone. Length 22 cm. Summer and winter plumage different. In winter upper parts black-brown and under parts white with broad dark band on breast; in summer upper parts tortoiseshell and head white. Short black bill; orange legs. Sexes alike. Call: up to 8 fast, low, slurred notes. On sandy and rocky or stony shores. Breeds mostly in the Arctic. Atlantic, English Channel, North Sea and Baltic. Common. (p. 88)

Numenius arquata (Linnaeus): Curlew. Length 55 cm. Grey-brown bird with long legs and very long bill curved downwards; white rump; under parts striped. No crown pattern on head. Sexes alike. Call: very distinctive 'coor-lee' with a mellow tone. Nests on moors, marshes and fields and feeds on mudflats, particularly in winter. Atlantic, English Channel, North Sea, Baltic and Mediterranean (winter only). Common. (**VI, 4**)

Limosa limosa (Linnaeus): Black-tailed godwit. Length 41 cm. Brown-grey upper parts, light below in winter; head and breast chestnut, broad black band on end of white tail in summer. Straight bill; broad white bar on wings in flight. Sexes alike. Call: 'wicka-wicka-wicka'. On mudflats and marshes on English Channel and Atlantic and Mediterranean coasts in winter; on Baltic, North Sea and Iceland shores and inland in summer. Apparently increasing in numbers. (p. 89)

Tringa totanus (Linnaeus): Redshank. Length 28 cm. Grey-brown upper parts with darker markings; light under parts;

white rump and white on tail edge of wings seen in flight; characteristic long red legs. Sexes alike. Erratic flight. Call: a 3-note whistle 'tew-tew-tew'; yodelling call when breeding. Often in large flocks on mudflats and open shores. Breeds inland in marshy areas. Atlantic, English Channel, North Sea, Baltic and Mediterranean (winter only). Common. (**VI, 17**)

Tringa nebularia (Gunnerus): Greenshank. Length 30 cm. Ash-grey upper parts with dark markings; under parts white with dark spots and bars in summer; wings unmarked; rump white. A long grey-blue bill and characteristic long green legs. Fast flight with rapid wing beats. Sexes alike. Call: a 3-note whistle. Nesting in northern Scandinavia and northern Scotland. Wintering in more southerly European coasts, down to Mediterranean and northern Adriatic shores. Atlantic north of English Channel, English Channel, North Sea and Mediterranean (winter only). Migrates over the Baltic to northern Scandinavia from Mediterranean in Spring. Common locally. (p. 89)

Calidris canutus (Linnaeus): Knot. Length 25 cm. Grey above and pale below in winter; mottled black and chestnut above and russet below in summer. Grey tail; light rump and wing-bar seen in flight. Dumpy and short-necked. Sexes alike. Call: a quiet 'wut' or whistling 'thu-thu'. Breeds in Arctic; seen on Atlantic, English Channel, North Sea and western Mediterranean coasts in winter in large flocks. Common. (p. 89)

Calidris alba (Pallas): Sanderling. Length 20 cm. In winter pale grey and white with black shoulder spot; in summer upper parts are brown with dark streaks. Sexes alike. Call: a sharp 'plick'. On sandy shores close to waterline. Atlantic, English Channel and North Sea in winter; nests in the Arctic. Common.

Calidris alpina (Linnaeus): Dunlin. Length 18 cm. Brown-grey streaked upper parts and white under parts (grey in winter); black belly in summer; white wing-bar and sides of tail seen in flight; bill curved slightly downwards. Sexes alike. Often seen flying in flocks. Call: a weak 'tweep' in flight. Nests in marshes inland but spends rest of year on mudflats and shores. Atlantic, English Channel, North Sea, southern Baltic and Mediterranean (winter only). Common. (**VI, 22**)

Larus ridibundus (Linnaeus): Black-headed gull. Length 20 cm. Chocolate-brown hood which is lost in winter; grey back and wings; black wing tips; rest of plumage white; red feet and bill. Call: a harsh cackle. Often seen inland as well as in estuaries and open coasts. Atlantic, English Channel, North Sea, Baltic and Mediterranean. Very common. (**VI, 1**)

Corvus corone cornix Linnaeus: Hooded or grey crow. Length 46 cm. Black with grey back and under parts. Sexes alike. Nests singly in trees, on cliff ledges or among heather. Often seen near rocky cliffs and near the shore. Northern and western areas of Britain and Ireland, Scandinavia and the Baltic area. Common. (p. 89)

Anthus spinoletta (Linnaeus): Rock pipit. Length 16 cm. Grey-brown upper parts. Grey under parts with dark broken streaks on breast. Grey outer feathers on tail. Dark brown legs. Sexes alike. Call: a single 'pheet'. On rocky shores in summer but may move to estuaries and saltmarshes in winter. Atlantic, English Channel, North Sea, Baltic and western Mediterranean (winter only). Common. (**II, 17**)

Motacilla alba alba Linnaeus: White wagtail. Black and white plumage with grey back; long tail constantly wagged up and down. Call: 'tchizzik'. Nests in holes in walls, banks and thatch. Runs fast, catching insects near the ground or occasionally in flight. Often on the shore. The pied wagtail (*Motacilla alba yarrellii*) is very similar in appearance, but is distinguished by the unbroken black of the 'bib' on the breast joining with the back of the head. In summer the male's back is

Motacilla alba alba: White wagtail

Motacilla alba yarrellii:
Pied wagtail

Halichoerus grypus: Grey seal

Phoca vitulina: Common seal

black and the female's is dark grey; both sexes are greyer in winter. Both species of wagtail are common. The pied wagtail is mostly limited to the British Isles, but it may occasionally be seen in Norway, Germany, Holland, Belgium and north-west France; the white wagtail occurs throughout north-west Europe. (p. 92)

Plectrophenax nivalis (Linnaeus): Snow bunting. Length 16 cm. In winter, has light brown back and crown of the head, with black and white wings and tail, and white breast. Looks very white when seen in flight, usually in flocks. In summer, the male's back becomes black. Call: 'tsweet' or 'teu'. Coasts of the English Channel northwards along western British Isles, North Sea and southern Baltic. Nests northern Scandinavia, northern Scotland and Iceland, usually in mountains. May be locally numerous in winter flocks.

Class: Mammalia (Mammals)

Halichoerus grypus (Fabricius): Grey seal. Very variable colour; spots where distinguishable are larger and less numerous than in common seal. Large head and high muzzle giving the head a horse-like appearance. Males up to 2·45 m long, females up to 2·2 m. On exposed rocky coasts and in caves; also in estuaries and on sandbanks. Large breeding aggregations are usually on offshore islands. Atlantic, North Sea and Baltic. Not uncommon in isolated areas. (p. 92)

Phoca vitulina Linnaeus: Common seal. Variable colour and pattern, usually mottle of dark spots on a lighter background. Forehead domed; face puppy-like. Males up to 1·85 m long, females up to 1·75 m. Usually found in shallow sheltered waters, sea lochs and around islands; often on sandbanks or mudbanks and in estuaries. Atlantic, south to Biscay, English Channel (occasionally), North Sea and Baltic. Less common than the grey seal. (p. 92)

CHAPTER TWO

The Underwater World

Equipped with modern diving gear and properly trained in its use, the marine naturalist is able to discover for himself the little-known world beneath the waves. He can only dip below the surface of the oceans, relatively speaking, but there is ample reward for doing so. A great wealth of marine life exists in the shallow seas of the continental shelf, and the era of its exploration is still so young that amateurs can expect to contribute to new knowledge by making original observations and recording what they see.

There is no space here to go into the techniques of diving and the equipment used. This is covered by the excellent manual of the British Sub-Aqua Club, which is recognised internationally as the standard work on the subject. Anyone intending to start Scuba diving, either as a sport in its own right or as a means of studying underwater life in its environment, cannot be recommended too strongly to receive training, either at a reputable diving school or with one of the diving clubs recognised by the World Underwater Federation (C.M.A.S.).

Although snorkel diving does not involve the sophisticated equipment required for Scuba diving, nor expose the diver to the attendant medical hazards, it is nevertheless an activity which should not be undertaken lightly. Again, anyone intending to take up the sport is urged to seek professional tuition, and familiarise himself with safety procedures.

Performing any kind of task under water is difficult. The time available is limited by the air supply and the diver's resistance to cold. A considerable part of even an experienced diver's concentration is devoted to the problems of staying alive in a hostile environment, and so the time available for attending to other tasks is further reduced. The underwater naturalist should start by making simple observations, taking notes and making sketches on a writing slate. This is easily made from white formica or a perspex sheet which has been roughened with fine sandpaper; an ordinary lead pencil can be used, and an india rubber will be a useful addition. A swimboard, similar to that used by Royal Navy divers, can be constructed to combine an instrument panel and writing slate (see Fig. 4).

Surveys can be carried out to estimate the abundance of animals or plants and how this changes with depth or perhaps substratum, using quadrats made from galvanised wire—or, on a larger scale, the method shown in Fig. 5. Divers are particularly well positioned to observe the behaviour of fish and other marine forms in perfectly natural conditions, and underwater photographs taken by divers combined with notes taken on the bottom are often far more instructive than photographs taken with cameras lowered from the surface. Underwater photography (dealt with in Chapter Five) is becoming increasingly popular with

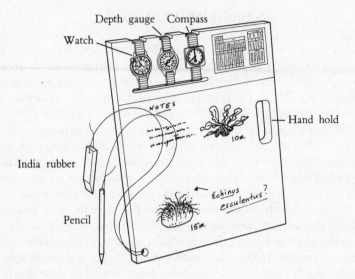

Fig. 4. Diver's swim board.

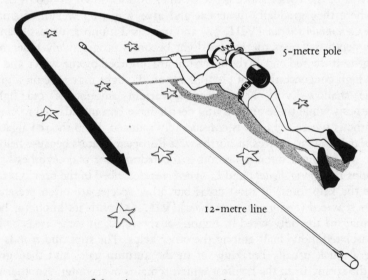

Fig. 5. Counting starfish using a 5-metre pole.

divers, justifiably so. It provides the ideal way of showing non-divers the beauty and attraction of the underwater world, and is a very acceptable alternative to collecting for the naturalist. In any case, avoid collecting marine life as far as possible, and on no account interfere with slow-growing fixed forms like sea-fans and cup corals. In this context also, do not get in the way of the work of professional inshore fishermen.

UNDERWATER LIFE

The snorkel diver need swim no further than the intertidal zone at high tide. When covered by water this habitat becomes transformed—all the fucoid algae, with their gas-filled flotation bladders, stand upright, exposing as great an area of their fronds as possible to the sunlight, and in this meadow of weed, crabs and prawns scuttle around in search of food, and different snails crawl over the rocks grazing the small algae or attacking their prey. Sea-anemones and hydroids extend their tentacles and filter-feeding bivalves and sea-squirts are active, pumping water through their filtration apparatus while the opportunity lasts. Small shore fishes leave the shelter of their rock pools which are now flooded and larger fish move in from the sublittoral zone to exploit the rich resources that this area has to offer. Sandy shores too come to life at high tide; bivalve siphons appear above the surface, shrimps and crabs leave their dugouts in the sand and the burrowing anemones push their tentacles up above the surface.

The fully equipped Scuba diver with aqualung and wetsuit is able to reach much greater depths, and he can stay down long enough to view this fascinating world at his leisure. The kelps or strap-weeds which extend upwards to the lowest levels of the rocky shore grow in thick forests down to depths of 10–15 metres where they gradually disappear and give way to a wealth of small red algae like *Ceramium rubrum* (**VIII, 32**) and increased numbers of sessile animals. As water depth increases the natural light becomes progressively bluer because wavelengths at the red end of the spectrum are absorbed by the water and where there are high concentrations of phytoplankton the light may take on a greenish tint. The shallow-living kelps need a certain amount of red light for photosynthesis while those algae living deeper have become adapted to carry out photosynthesis where red light is considerably reduced. Even the red algae begin to disappear below 30 metres in north-west European waters because below this depth the overall light intensity becomes too reduced for photosynthesis.

The kelps *Laminaria digitata* and *L. hyperborea* described in the previous chapter dominate the kelp forest in most areas but other species are often present. The furbelows seaweed (*Saccorhiza polyschides*) (**VII, 15**) with its knobbly, bulbous holdfast may be the only weed in regions of the forest in some areas or it may occur as isolated individuals among the other kelps. The stipe and fronds of this weed are annual, usually breaking off in the autumn gales and then growing again in the spring from the holdfast which remains overwinter. Another species of kelp usually found in sheltered conditions is the sugar kelp or poor man's weather glass (*Laminaria saccharina*) (**VII, 19**). It is fond of areas where there is a sandy or gravel bottom and under these circumstances will attach itself to a small stone or pebble. In areas where it grows in high densities, such as off the beach at Worthing on the south coast of England, it may play an important role in the transport of pebbles. During stormy weather when there is a lot of turbulence the large frond acts as a drogue and lifts the pebble to which it is attached off the bottom to be deposited elsewhere.

There are several other brown weeds which grow amongst kelp. Thong weed (*Himanthalia elongata*) (**VII, 23**) has a holdfast which attaches it to the rock surface, and immediately above that there is a button-like structure resembling an enlarged golfer's tee-peg. This structure is sometimes all that is present but usually the long, branched, strap-like thong grows up from its centre. In sheltered situations and usually where there is a little sand or gravel the bootlace weed (*Chorda filum*) (**VII, 17**) grows in dense strands. In shallow places at low water the upper portions of the air-filled floating strands may lie on the surface of the sea.

Although the brown seaweeds dominate the seabed in shallow water a great many other algae are also to be found there. The green algae are largely restricted to the shore but in shallow water, particularly where there is mud and sand among the rocks, the sponge seaweed (*Codium tomentosum*) (**VII, 16**) may grow, and where this is present the small, slug-like, saccoglossan mollusc *Elysia viridis* (**VII, 18**) will almost certainly be found. This animal usually grazes on *Codium* and other green algae such as *Ulva* and *Enteromorpha* which it slits open with its scalpel-like radula. It then sucks out the juices by a pumping action of the pharynx. The colour of this mollusc is determined by the algae upon which it is feeding, and so it is usually green with a number of glistening white spots. If its diet is changed to a red species of algae it will take on a reddish hue within about three weeks.

Around the base of the kelp forest, among the holdfasts, there is a considerable undergrowth of small red seaweeds. *Dilsea carnosa* (**VIII, 15**) and *Delesseria sanguinea* (**VIII, 16**) may be quite conspicuous, producing large leaf-like growths, while other species like Irish moss or carragheen (*Chondrus crispus*) (**VII, 4**) form a carpet-like covering over the rock.

Although there is a great deal of attached weed growth in shallow water where the bottom is rocky, there are comparatively few grazing animals both in terms of variety and total numbers. This at first may seem surprising but it may be important for the economy of the seabed as we shall see later. The most conspicuous grazing animals of the kelp forest are the sea-urchins. The edible sea-urchin (*Echinus esculentus*) (**VII, 12**) occurs on most coasts, except in the Baltic and the eastern English Channel, from the extreme lower shore to depths of 50 metres, and is very easily spotted, being both large and brightly coloured. Where it is found below the level of seaweed growth it is probably browsing on encrusting and sessile animal life such as hydroids, barnacles and tube-dwelling worms. The purple urchin (*Paracentrotus lividus*) (**VIII, 31**) is less widely distributed but may be abundant in localised areas on the west of Ireland and the French Atlantic coast. It bores down into the rock on which it lives and sometimes becomes incarcerated by the growth of the calcareous, encrusting red algae *Lythophyllum* mentioned in the previous chapter. Under these circumstances it is not clear what the sea-urchin is feeding upon but there are suggestions that it is able to gain nutriment from the small amounts of dissolved organic matter which are present, particularly in coastal waters.

Sea-urchins have the mouth on the underside of the body and this is armed with a complex jaw mechanism known as the Aristotle's lantern. Five teeth which are moved by a set of rods and muscles are used to scrape algae and small animals from the rock as the animal moves over the surface, using the suckers on its extended tube feet for attachment and to pull itself along. Although sea-urchins may occur in large numbers their existence is threatened in some areas by excessive collection by divers.

The sea-hare (*Aplysia punctata*) (**VIII, 6**) eats algae and there are a few small grazing snails such as the painted topshell (*Calliostoma zizyphinum*) and grey topshell (*Gibbula cineraria*), described in the previous chapter. These are fairly ubiquitous but it is only in the shallow waters of the coast of Brittany and the Channel Islands that we find the green ormer (*Haliotis tuberculata*) (**VIII, 11**). This primitive limpet-like snail with its drab exterior and beautiful mother-of-pearl interior is prized by gourmets, and sadly has been severely reduced in numbers by overfishing. It is immediately recognisable by the border of small holes in the shell. It lives in thick kelp forest from the lower shore to around 30 metres but is most abundant at about 2–3 metres. During the day these animals remain beneath the rocks; at night they become active and move over the rocks, at times with surprising speed, to graze their favourite red and green algae.

Although these grazing animals are large and conspicuous their low density means that they only consume a small amount of the plant material produced, in comparison with how much terrestrial animals graze on grassland for instance, or even the zooplankton grazing on the phytoplankton. The lower level of grazing on the seabed, however, may be important in maintaining the ecological balance. On the land a grazing animal like a cow quickly recycles the nutrients it has eaten because its dung goes straight back onto the ground. On the seabed however the nutrients released in the faeces of a bottom-living grazing animal will be dissolved in seawater and will very likely be carried away so that they cannot be recycled by the plants in the community where they originated. Clearly, high grazing rates in this situation could be detrimental to the stability of the community.

It should also be pointed out that the seaweeds in general have little nutritional value as they are rather poor in protein. It is perhaps therefore not surprising that few animals have evolved to feed on them and some that do, like the sea-urchins, browse on a certain amount of animal food as well.

The kelp forest provides a habitat for a great many mobile animals: predatory echinoderms—the spiny starfish (*Marthasterias glacialis*) (**VII, 11**) and the cushion star (*Porania pulvillus*) (**VIII, 25**), beautiful nudibranch sea-slugs, squat-lobsters (*Galathea squamifera*) (**I, 10**), edible crabs (*Cancer pagurus*) (**VIII, 29**) with their brown pie-crust shell, and spiny spider crabs (*Maia squinado*) (**VIII, 3**).

Edible crabs abound in sandy patches and rocky weed-strewn areas from shallow waters. They generally hide under stones or in crevices. As scavengers, their diet embraces a wide variety of both dead and living fish and invertebrates. In late spring they migrate inshore where the ripe eggs which are carried by the female are hatched and where moulting and copulation occur. A few days prior to the female moulting a male is strongly attracted to her. He assumes a protective position astride the female, clasping her back against his ventral surface. If disturbed, the male becomes very aggressive and drives off any intruders. Occasionally though, he may be ousted by a larger male who will then take up the position of protector. When the female is moulting the male assists by helping her out of her shell and copulation follows when the moult is completed. In winter the crabs again move offshore where the eggs are extruded from late November to February, a female 15 cm wide extruding about one million eggs. The eggs are hatched during the summer the following year and the zoea described in the next chapter spend 20–30 days in the plankton before settling to the bottom as juveniles.

Although crabs are landed in most European countries, the most important crab-fishing areas in European waters are off Norway, around the coast of Scotland and along east and south-west England. Smaller fisheries occur off Spain and Portugal and around Ireland. Crabs are fished in baited traps known as pots and although these vary from place to place there are two main types: creels, which are cylindrical in shape, and inkwell pots whose name describes their general shape.

The spiny spider crab is generally found associated with sand or rock. Unlike the edible crab, spider crabs do not hide in crevices but rather rely on an unusual form of camouflage for concealment. All adult crabs of this species attach pieces of algae and sand grains to the dorsal surface of their carapace and these remain in place, so providing a degree of concealment. While attracted to baited pots, spider crabs are mainly vegetarian, feeding mostly on algae but also hydrozoans and polyzoans. Fish do not appear to form part of their natural diet.

In winter spider crabs migrate to offshore waters. However, in summer they may be seen in startlingly large numbers in shallow inshore areas. This inshore movement, commencing in late spring or early summer, occurs prior to moulting and breeding. In these inshore waters spider crabs arrange themselves into dome-shaped mounds. The centre of these contain immature males and females about to moult to maturity. Encapsulating these are large mature specimens, generally males which copulate with newly moulted females. About six months after copulation the eggs are extruded onto the abdominal appendages (pleopods) of the female. Up to 150,000 eggs may be extruded by a female at any one time. The eggs remain attached to the pleopods while developing and in late summer they hatch as zoea larvae which measure about 2·5 mm in length. The zoea spend about twenty days swimming in the plankton before they moult into the first crab-like benthic stage. There has been a commercial fishery for spider crabs in Europe for some years and since 1976 a fishery has been developing on the south coast of England to supply the French and Spanish markets.

Fishes are always to be found in the vicinity of the kelp forest. Ballan wrasse (*Labrus bergylta*) (**VII, 1**) swim close to the bottom and usually remain among the holdfasts. Occasionally one may come across a specimen lying on its side in a

patch of gravel and apparently asleep. In early summer the male and female build
a nest of seaweeds, particularly the coralline algae, torn from the rocks. This is
packed into a rock crevice and the eggs are laid inside. Pollack (*Pollachius
pollachius*) (described more fully in the next chapter) swim in large shoals over the
roof of the kelp forest and will feed on sand-eels which suddenly shoot up from
sandy patches if the bottom is disturbed by a diver. The lesser spotted dogfish
(*Scyliorhinus caniculus*) (**VII, 7**) seems to spend a great deal of time lying idly
among the holdfasts but it may be seen swimming slowly along the bottom using
its well-developed sense of smell to search out food. The mouth of this fish is
designed for feeding on bottom-living animals and it normally takes such prey as
small crabs and other shellfish.

The lesser or curled octopus (*Eledone cirrhosa*) (**VII, 25**) is not uncommon in
waters off Atlantic coasts, in the English Channel and the northern parts of the
North Sea. The common octopus (*Octopus vulgaris*) is more restricted, usually
extending only from more southerly areas as far as the western part of the English
Channel. Both species are usually extremely difficult to spot as they camouflage
themselves to match their surroundings by expanding and contracting the
pigment cells in their skin. They are carnivores, feeding on crustaceans and other
small prey. A crab is captured by a process of stalking followed by a sudden
pounce, after which it is killed by biting with the parrot-like beak and injecting
with a poison. Digestive enzymes are then shed into the shell of the crab which is
kept until a soup-like mass has been produced. The soup is then sucked out and
the more or less intact crab shell discarded.

Beyond the lower limits of the kelp forest and in caves, where the absence of
light prevents plant growth developing, all rock surfaces become colonised by
fixed or sessile animals which encrust and adorn the surface to the extent that it is
usually impossible to find a bare patch of rock. Some sponges like the black
elephant's ear sponge (*Pachymatisma johnstoni*) may cover over a square metre of
rock, and hydroids, sea-anemones, fan worms and sea-squirts abound. The horse
mussel (*Modiolus modiolus*) (**VIII, 20**) sometimes occurs in densely packed clumps
of about 5 to 30 individuals, and these clumps create an environment which
attracts a wealth of mobile species including small crabs, whelks, starfish, brittle
stars and sea-urchins plus a variety of sessile forms like tube worms, barnacles,
and sea-squirts.

In many coastal situations where there is a rocky shore this continues down
into the sublittoral zone for a relatively short distance, giving way to a softer
substrate which may be muddy in enclosed sites such as sea lochs or sandy gravel
where conditions are more exposed. In a few places in the west of Ireland and off
Brittany the rock gives way to what is commonly termed a 'coral' bottom
although in fact the 'coral' is a species of the branching calcareous algae,
Lithothamnion or maerl, closely related to the pink encrusting *Lithophyllum
incrustans* described in Chapter One.

These various types of bottom usually tend to slope shallowly towards the open
sea and they support little or no growth of weed. There are sometimes rocky

outcrops however and in shallow water these will provide a site for attachment of kelp while in deeper water they will tend to have a covering of sessile animals and sea-urchins. The submarine deposits have a far greater number of epifaunal species, or species which do not burrow, than their counterpart in the intertidal zone. Brittle stars may form dense aggregations and various starfish—*Porania pulvillus* (**VIII, 25**), *Henricia sanguinolenta* (**VIII, 28**), the common sunstar (*Crossaster papposus*) (**VII, 28**), the spiny starfish (*Marthasterias glacialis*), mentioned earlier, and the common starfish (*Asterias rubens*), mentioned in the previous chapter—sit on the bottom and prey on bivalves which they locate by their sense of smell.

Beds of variegated scallops (*Chlamys varia*) (**VIII, 7**), queen scallops (*Chlamys opercularis*) (**VIII, 13**) and great scallops (*Pecten maximus*) (**VII, 26**) are often found lying on the substratum in deeper water close to the shore. These scallops, particularly queens and the great scallop, are able to swim strongly for short distances when threatened by a predator, such as a starfish. They can detect a potential predator by the equivalent of their sense of smell, when they are physically disturbed, or sometimes when a passing shadow is detected by the eyes which line the mantle edge. Swimming is achieved by the scallop clapping the valves of the shell together, forcing a jet of water out through openings near the 'ears'. C. M. Yonge describes the action as 'the animal moving by taking bites out of the water ahead'. The great scallop uses a similar method to turn itself over if placed upside-down, and the variegated scallop, despite the fact that it attaches to the bottom by the byssus, is reported to release itself and start swimming when threatened. In addition to this line of defence the variegated scallop often has a protective covering in the form of the crumb-of-bread sponge (*Halichondria panicea*), described in the previous chapter, which grows on the shell. The sponge prevents starfishes gaining a firm hold on the shell with their tube feet and so they are unable to prise the two valves apart. In return the sponge, which is a suspension feeder, probably benefits from the water current set up by the scallop for its own respiratory and feeding purposes.

A variety of fishes occurs on the seabed below the level of the kelp forest. Many of these are dealt with in the next chapter but there are a couple which divers are particularly likely to see at close quarters and should be mentioned here. The dragonet (*Callionymus lyra*) (**VII, 24**), is common in all areas; it has a flattened underside, which suits its bottom-living habit, and a rounded back. The pectoral fins are enlarged and these enable them to make sudden darts forward to catch their prey which consists of bottom-living crustaceans, molluscs and worms. In spring the male dragonet takes on his full breeding coloration in which his yellow-brown body becomes spotted with a lustrous azure blue, the fins take on a darker blue and the eyes become brilliant blue-green. The male displays before a female at this time with his blue and yellow dorsal fin erect and lips pursed. If the female should accept him the male lifts her gently on his pelvic fins and they swim side by side almost vertically towards the surface, shedding their eggs and sperm as they go.

The thornback ray (*Raja clavata*) (**VII, 22**) is extremely well camouflaged against the gravel bottoms on which it usually lies and divers frequently approach one of these large fishes unaware of its presence. When disturbed the ray will leap off the bottom in a great flurry and glide quickly away to settle down again after swimming a few metres.

Submarine deposits support a rich infauna of burrowing animals. Many of these spend most of the time either at the entrance to their burrow or scuttling around over the surface, so the diver has an ideal opportunity to observe them without disturbance. The anemone-like *Cerianthus lloydi* inhabits a slimy tube in the mud and withdraws with a rapid jerk if approached closely. Sea-pens (*Pennatula phosphorea*) (**VIII, 14**) extend well above the surface of soft muddy bottoms and they too construct a mucus-lined burrow and can withdraw fully if disturbed. They use stinging cells of the same type as the sea-anemone's to capture small prey and they may also carry out a form of suspension feeding. Soft bottoms provide a habitat for a great many bivalves: some such as the dog cockle (*Glycymeris glycymeris*) (**VIII, 4**) are restricted to the sublittoral region but many species from the seashore will also be met by the diver.

In deep water off Atlantic and North Sea coasts where the bottom is muddy the diver is likely to meet the Norway lobster (*Nephrops norvegicus*) (**VIII, 23**), otherwise known as the Dublin Bay prawn or scampi. *Nephrops* inhabits U-shape burrows and if these are approached quietly the inhabitant is usually seen at the entrance with antennae and claws protruding. They seem to leave their burrows at dawn and dusk and it is at these times that most are caught by fishermen with trawl nets.

Shipwrecks always have an uncanny fascination for divers. In shallow water, wrecks quickly become broken up by winter storms and the remains are then rapidly concealed by weed growth which often renders them unrecognisable. In deeper water, however, wrecks remain intact for longer and seem to have a magnetic attraction for marine life as well as human explorers. If the wreck is deep enough for there to be insufficient light for plant growth the plates of a steel ship soon become covered with huge growths of the beautiful pink and white dead man's fingers (*Alcyonium digitatum*) (**VII, 29**).

When a wreck becomes broken up it creates an artificial reef on the seabed to which fishes are greatly attracted. They create good fishing marks for anglers and are well worth a visit by divers. Huge shoals of pout or whiting (*Trisopeterus luscus*) (**VII, 8**) and Ballan wrasse (*Labrus bergylta*) (**VII, 1**) swim among the wreckage, and ling (*Molva molva*) (**VII, 6**) and conger (*Conger conger*) lurk in the many nooks and crannies. Often it is only the head and neck of these last two fishes that can be seen when they are in their lairs, but a single barbel on the chin identifies the ling, which also has a well-developed gill cover.

Some of the most spectacular diving off the coasts of north-west Europe is to be had near the rocky offshore islands of north-west Scotland, western Ireland, south-west England and Brittany. Away from the effects of the land one finds clear blue oceanic water and several species of animal which are only to be found where no influence of coastal water exists. Diving on the edge of the dropoff from these islands, steep rocky cliffs and broad ledges are seen to be covered with magnificent growths of sea-fans (*Eunicella verrucosa*) (**VIII, 26**), rose coral (*Pentapora foliacea*) and large areas of rock may be obscured by aggregations of the rosy feather star (*Antedon bifida*) (**VIII, 27**). The feather stars feed on organic particles suspended in the water but unlike many other suspension feeders they are unable to create a water current themselves. They inhabit places where gentle water currents are present and they hold their tube feet across the direction of the current to trap particles as they are carried past. The jewel anemone (*Corynactis viridis*) (**VIII, 19**) covers offshore rocks in spectacular colours of green and pink, and the brilliant white anemone *Actinothoë sphyrodeta* (**VIII, 18**) is usually present on vertical and overhanging rock faces and on cave walls. Here too lives the solitary Devonshire cup coral (*Caryophyllia smithi*) (**VII, 5**) which shows densities of over 100 per square metre on clean rocky surfaces. It is very slow-growing and feeds, like the anemones, on small crustaceans and other tiny animals found near the rock face.

Crevices in the cliff face often provide a home for the common lobster (*Homarus gammarus*) (**VII, 10**). This much sought after crustacean occurs wherever there are rocks and is most frequent between depths of 20–30 metres, although they occasionally turn up in shallower water and even between tide marks. They are solitary animals except in the breeding season when a mating pair may share a crevice for a few days. Lobsters are mostly active at night when they leave their hole or cave to feed on a variety of fish and other invertebrates— dead or alive.

The spiny lobster (*Palinurus vulgaris*) (**VIII, 24**) commonly occurs on rocky reefs in deep waters of about 30–40 metres, although rarely they are found in waters 12–15 metres deep. They are nocturnal animals and shelter in caves and crevices during the day. They are gregarious and often groups of up to ten animals may inhabit the same shelter although they generally seek separate homes. Spiny lobsters feed on a large variety of marine animals but echinoderms and molluscs form the main part of their diet. They exhibit two migrations annually; an onshore movement in late spring or early summer, and an offshore move in winter to deeper waters. They are often found migrating together in large numbers at densities up to 40 per square metre and congregations of spiny lobsters are often seen on reefs in the summer. When examined more closely they are usually found to be females attracted to the area to extrude their eggs. If disturbed, spiny lobsters will let out an extraordinarily loud creaking noise which they produce by stridulation similar to the way their distant cousins, the grasshoppers and crickets, produce their characteristic song. They do this by scraping a file-like structure on the base of the antennae against the rough surface of a sound box beneath the rostrum (grasshoppers and crickets scrape the inside of the legs against the side of the body).

Fish are usually plentiful at offshore sites; the brilliant colours of the cuckoo wrasse (*Labrus mixtus*) (**VII, 2, 3**) show up to their best in crystal clear water, and it is offshore that one stands the best chance of seeing shark. There has never been a recorded attack by shark in our waters but one would be advised to be at least wary of those species (described in the next chapter) like the mako (*Isurus oxyrinchus*) (**XI, 2**), which are known to have been responsible for attacks elsewhere. The basking shark (*Cetorhinus maximus*) (**XI, 1**) is completely harmless, despite its large size, and it is a most thrilling experience to be in the water with these fish; however close they come towards a diver they have an unerring ability to avoid even the slightest contact.

Diving at night can be a wonderful experience for the naturalist. After sunset, armies of crabs and other crustaceans, which remain hidden in their lairs during the day, leave their shelter and may be seen scuttling over the bottom in search of food. Fish become mesmerised by the light of a diver's torch and can be approached far more easily than during the day; and if the torch is turned off luminescent planktonic organisms appear as explosions of light in the wake of one's fins and with every movement.

PLANTS AND ANIMALS OF THE UNDERWATER WORLD

Plant Kingdom

Division: Algae
Class: Chlorophycae (Green algae)

Codium tomentosum Stackhouse: Sponge seaweed. Dark green frond 25–35 cm high. Tubular fleshy threads which branch dichotomously; felt-like texture; disc-like holdfast. Growing on rocks, mud and sand down to 20 m. Atlantic, English Channel and Mediterranean. Common locally. (**VII, 16**)

Class: Phaeophycae (Brown algae)

Saccorhiza polyschides (Lightfoot): Furbelows seaweed. Holdfast a large, bulbous, hollow structure with knobbly surface; flattened stipe and massive fronds. Whole plant may be over 2 m high. Stipe and fronds are annual. Extreme lower shore to 10 m. Atlantic, English Channel and North Sea. Common. (**VII, 15**)

Laminaria saccharina Lamouroux: Sugar kelp or poor man's weather glass. Brown thallus with crumpled appearance up to 3 m long; no midrib. Round stalk, relatively thin and up to $^1/_4$ the length of the blade. Branching holdfast. Middle shore to shallow water; often attached to stones on mudflats and sandy bottoms. Atlantic, English Channel and North Sea. Common. (**VII, 19**)

Chorda filum (Linnaeus): Bootlace weed. Slippery, string-like weed 2–6 m in length. Brown, unbranched and round in section. Small disc-shaped holdfast. Lower shore to 20 m, usually in shallow water. Atlantic, English Channel, North Sea and Baltic. Common. (**VII, 17**)

Himanthalia elongata (Linnaeus) S. F. Gray: Thong weed. Dichotomously branched, brown, strap-like frond arising from flat-topped mushroom-like base. Up to 2 m long. Frond may be absent, leaving mushroom-like structure. Lower shore and shallow water on exposed coasts. Atlantic

and English Channel. Common locally. (**VII, 23**)

Padina pavonia (Linnaeus) Lamouroux: Peacock's tail. Fan-shaped frond 4–10 cm long; yellowish-green with concentric silver markings, darker towards the base. Shallow water, attached to rocks. Atlantic, English Channel and Mediterranean. Common locally. (**VII, 9**)

Class: Rhodophyceae (Red algae)

Chondrus crispus Stackhouse: Irish moss or carragheen. Very variable seaweed, usually a flat frond, dichotomously branched with a short stalk. Dark red, may possess a violet iridescence under water. Fruiting bodies appear as small swellings on the stem in winter. Lower shore and shallow water, attached to stones and rocks. Atlantic, English Channel, North Sea and Baltic. Common. (**VII, 4**)

Lithothamnion calcareum (Pallas): Maerl. May occur as a flat, encrusting pink growth but often raised into knobby branches forming 'heads' up to 8 cm in diameter. In certain areas the calcareous skeletons form extensive bottom deposits. Extreme lower shore and shallow water; often unattached. Atlantic, English Channel and North Sea. Common locally.

Dilsea carnosa (Schmidel) Kuntze. Dark red, very tough weed up to 30 cm long. Short rounded stipe attached by disc-shaped holdfast. Frond may be much cut by wave action. Lower shore to shallow water on rocks, stones or on sandy bottoms. Atlantic, English Channel and North Sea. Common. (**VIII, 15**)

Ceramium rubrum (Hudson) C. A. Agardh. Bushy seaweed with springy texture and dichotomously branched stems, forcipate tips. Red with fine banding. Up to 30 cm long; very small holdfast. Middle shore to deep water, attached to rocks and other seaweeds. Atlantic, English Channel, North Sea, Baltic and Mediterranean. Common. (**VIII, 32**)

Delesseria   sanguinea   (Hudson) Lamouroux. Thallus consists of a branched stalk bearing large leaves with a midrib and wavy, undivided margins. Deep pink. Lower shore and shallow water in the kelp forest. Atlantic, English Channel, North Sea and Baltic. Common. (**VIII, 16**)

Animal Kingdom

Phylum: Porifera (Sponges)

Suberites domuncula (Olivi): Sea-orange. Orange-yellow sponge growing in globular, brain-like shapes up to 20 cm in diameter; no oscula. May encrust rocks or piles and is often found on shells inhabited by hermit crabs. From shallow to deep water. Atlantic, English Channel, North Sea and Mediterranean. Not uncommon. (**VIII, 22**)

Pachymatisma johnstoni (Bowerbank): Elephant's ear sponge. Massive sponge, up to 60 cm across; dark grey with large oscula. From extreme lower shore downwards; under rocky overhangs and caves in areas exposed to wave action. Atlantic, English Channel and North Sea. Common.

Phylum: Coelenterata (Sea-firs, Sea-anemones, Corals, Jellyfish)
Class: Hydrozoa (Sea-firs)

Plumularia catharina Johnston. Stem up to 10 cm high with side branches arranged opposite one another. Deep thecae borne on stem and branches. Lower shore and below; attached to stones, shells and sea-squirts. Atlantic, English Channel and North Sea. Common. (**VIII, 21**)

Class: Anthozoa (Sea-anemones, Corals, etc.)

Cerianthus lloydi Gosse. Polyp up to 20 cm long with 60–70 brown peripheral tentacles up to 4 cm long and a similar number of small inner tentacles. Body yellowish-white, red or brown; tentacles brown with white rings at the tips. Inhabits a mucous tube buried in muddy sand at depths of 1–35 m. Atlantic, English Channel and North Sea. Not uncommon.

Anemonia sulcata (Pennant): Snakelocks anemone. Up to 12 cm across and 5 cm high. 180–200 long, blunt, non-rectractile tentacles arranged closely in a number of circles. Slightly adhesive base; smooth column. Green grey or brown often with purple tips to the tentacles. From lower shore down to 20 m, attached to rocks and seaweeds; usually in well-lit areas. Atlantic, English Channel and Mediterranean. Common. (**VII, 20**)

Metridium senile (Linnaeus): Plumose anemone. Up to 30 cm high and 20 cm across. Mouth disc lobed with up to 1000 fine tentacles giving a feathery appearance. Colour usually pink or white. On rocks and man-made structures down to 3 m. Atlantic, English Channel, North Sea and Mediterranean. Common. (**VIII, 17**)

Adamsia palliata (Bohadsch): Cloak anemone. Always found on shells inhabited by the hermit crab *Eupagurus prideuxi*. Column and base modified to attach to shell. 500 tentacles directed downwards in 8 alternating circles. Body brown or yellow with red spots; tentacles white. 4–100 m, on mud and sand. Atlantic, English Channel, North Sea and Mediterranean. Common. (**VII, 13**)

Actinothoë sphyrodeta (Gosse). Base up to 1 cm across, column up to 3 cm high, 100 white tentacles. On rock faces and in caves, usually in deeper water. Atlantic and western English Channel. Common. (**VIII, 18**)

Corynactis viridis Allman: Jewel anemone. Small polyp up to 0·5 cm in diameter with 3 circles of knobbed tentacles. Brilliantly coloured; often pink and green forms occur together in dense populations on rock faces. Down to 100 m, usually in clean waters. Atlantic north to south-west Ireland. Common locally. **(VIII, 19)**

Caryophyllia smithi Stokes: Devonshire cup coral. Polyp white or pink and up to 1·5 cm high; small knobbed tentacles. Skeleton white with radiating septa. Solitary. Extreme low water to 100 m, attached to rocks. Atlantic, English Channel, North Sea. Common locally. **(VII, 5)**

Eunicella verrucosa (Pallas): Sea-fan. Colony branching in one plane only. Horny skeleton, 30 cm or more high, covered by pink tissue in which lie the numerous polyps. Firmly attached by the holdfast to rocks and boulders at depths from 15 m. Atlantic, western English Channel and .Mediterranean. Common locally. **(VIII, 26)**

Alcyonium digitatum (Linnaeus): Dead man's fingers. Upright, bulbous colonies, usually branched and up to 20 cm or more in height. White, yellow, orange or pink with iridescent white polyps up to 1 cm long. Colony contracts when touched. On rocks and boulders, down to 100 m. Atlantic, English Channel and North Sea. Common. **(VII, 29)**

Pennatula phosphorea (Linnaeus): Phosphorescent sea-pen. Feathery colony on a slender stalk up to 40 cm high. Red, pink or white with small white polyps. From 20 m downwards, on mud and clay. Atlantic, North Sea, western Baltic and Mediterranean. Not uncommon. **(VIII, 14)**

Phylum: Annelida (Segmented worms)
Class: Polychaeta (Bristle worms)
Bispira volutacornis (Montagu). Fan

consists of 2 twisted gill clusters with eye-bearing filaments. Round body up to 15 cm long with reduced head and about 100 segments. Inhabits membranous tube covered with mud. Shallow water and downwards, attached to underside of rocks. Mediterranean, Atlantic, English Channel. Not common. **(VIII, 12)**

Phylum: Mollusca (Molluscs)
Class: Gastropoda (Snails and slugs)
Haliotis tuberculata Linnaeus: Green ormer. Flattened spiral shell with a series of openings along the top; greenish-brown outside, heavy mother-of-pearl inside; up to 8 cm long. On rocks and boulders at extreme lower shore and in shallow water. Atlantic northwards to the Channel Isles and Mediterranean. Not common due to over-exploitation. **(VIII, 11)**

Gibbula magus (Linnaeus): Topshell. Flattened shell possessing 8 whorls with conspicuous ridge at base; large umbilicus. Up to 2·5 cm in diameter. Yellowish-white shell with red markings. Extreme lower shore and below; buried in sand. Atlantic, English Channel, North Sea and Mediterranean. Not uncommon. **(VIII, 8)**

Clathrus clathrus (Linnaeus): Common wentletrap. Tall shell up to 4·5 cm high. 15 whorls with conspicuous diagonal ridges. Whitish, brown or reddish in colour. Shallow water down to 80 m, on sand and mud. Atlantic, English Channel, North Sea, western Baltic and Mediterranean. Not uncommon. **(VIII, 9)**

Trivia monacha (da Costa): European cowrie. Shell up to 1·2 cm long with ridged slit-like opening; fine ribs on upper surface. Pinkish brown with 4–6 dark brown spots. When undisturbed yellow-orange mantle lobes cover the shell. Lower shore and shallow water; may be found feeding on compound ascidians such as *Botryllus*. Atlantic, English Channel, North Sea and Mediterranean. Not uncommon. **(VIII, 10)**

Aplysia punctata Cuvier: Sea-hare. Up to

15 cm long; body almost encloses a much reduced shell which is thin and translucent. 2 pairs of head tentacles, the posterior pair resembling ears. Young individuals are red but turn brown or green with age. A red dye is released if disturbed. In shallow water among *Laminaria*; may be found onshore in summer when spawning. Atlantic, English Channel, North Sea. Common; occasionally occurs in large numbers. (**VIII, 6**)

Elysia viridis (Montagu). Shell lacking, body up to 3 cm long, soft and flattened with lateral lobes; 2 head tentacles. Bright green with white spots. Middle shore and into shallow water usually on green seaweeds such as *Codium*. Atlantic, English Channel, North Sea, western Baltic and Mediterranean. Not uncommon. (**VII, 18**)

Limacia clavigera (O. F. Müller). Flat white body about 2 cm long possessing over 20 appendages with orange-red tips. 3 branched gills around the anus. No shell on shallow water. Atlantic, English Channel, North Sea and Mediterranean. Not uncommon. (**VIII, 2**)

Facelina auriculata (O. F. Müller). Thin body but up to 2·5 cm long with 2 dissimilar pairs of head tentacles. About 6 groups of red body appendages with white tips. No shell. Lower shore and shallow water. Atlantic and English Channel. Common. (**VIII, 5**)

Class: Bivalvia (Bivalves)

Glycymeris glycymeris (Linnaeus): Dog cockle. Rounded shell up to 8 cm long, similar valves with crenulate edges and 2 rows of about 12 teeth on each hinge. Light brown with dark brownish-red zig-zag markings outside, white with brown patch inside. From shallow water to 80 m, in sand and mud. Atlantic, English Channel, Baltic and Mediterranean. Common. (**VIII, 4**)

Modiolus modiolus (Linnaeus): Horse mussel. Shell up to 20 cm long, oblong or irregularly triangular in outline with beaks a short distance from the anterior end. Dark blue or purple with glossy yellow or dark brown periostracum. In juveniles the periostracum is drawn out into smooth spines. Anterior adductor scar small, posterior large. Extreme lower shore to 150 m, often in large clumps. Atlantic, English Channel and North Sea. Common locally. (**VIII, 20**)

Pecten maximus (Linnaeus). Great scallop. Shell up to 15 cm long with equal ears. 1 valve convex and slightly overlapping the other valve which is flat and lies uppermost. 15–17 broad radiating ribs and numerous concentric lines. Flat valve brownish-red, convex valve white, cream or light brown. In deep water on stable sand and gravel. Atlantic, English Channel and North Sea. Common. (**VII, 26**)

Chlamys varia (Linnaeus): Variegated scallop. Shell up to 6 cm long or occasionally larger. Anterior ear 2–3 times longer than posterior; oval in outline; both valves convex; 25–35 radiating ribs. Variable colour: red, white, pink, brown or any combination of these. From extreme lower shore to 80 m, attached to rocks by the bysuss or on oyster beds. Atlantic, English Channel, North Sea and Mediterranean. Common. (**VIII, 7**)

Chlamys opercularis (Linnaeus): Queen scallop. Shell up to 9 cm long and circular in outline. Ears almost equal; left or upper valve more convex than the right; 19–22 radiating ribs. Usually in deeper water on sand and gravel, juveniles attached by bysuss but adults free-living. Atlantic, English Channel, North Sea and Mediterranean. Common. (**VIII, 13**)

Class: Cephalopoda (Cuttlefish, octopus and squid)

Sepiola atlantica d'Orbigny: Little cuttlefish. Rounded body up to 5 cm long with lobe-like lateral fins. Variable colour which may change when disturbed. In shallow water on sand. Atlantic and English Channel. Common. (**VII, 14**)

Eledone cirrhosa Lamarck: Lesser or curled octopus. Total length up to 50 cm. Tentacles have a single row of suckers. Reddish-yellow with brown spots. Among rocks in shallow water. Atlantic, Channel and North Sea. Not uncommon in summer. (**VII, 25.**) The common octopus *Octopus vulgaris* has a double row of suckers on the tentacles and is usually only found as far north as the English Channel. It may be common after mild winters.

Phylum: Arthropoda (Arthropods)
Class: Crustacea

Homarus gammarus (Linnaeus): Common lobster. Body up to 45 cm long. Smooth carapace with mottled blue markings and orange tints on a white background; red antennae. 1st pair of walking legs possesses large pincers, a 'cutter' and a 'crusher'. From shallow water to 30 m in holes and caves. Atlantic, English Channel, North Sea, western Baltic and Mediterranean. Common. (**VII, 10**)

Palinurus vulgaris Latreille: Spiny lobster. Body up to 50 cm long. Spiny carapace coloured red or brown. No large pincers. Long antennae. From shallow water to 70 m in holes and caves. Atlantic, western English Channel and Mediterranean. Not uncommon. (**VIII, 24**)

Nephrops norvegicus (Linnaeus): Scampi, Norway lobster or Dublin Bay prawn. Body up to 15 cm long. Slender with long unequal pincers on 1st pair of walking legs. Long slender 2nd antennae. Salmon-pink. From 20 m downwards in 'U'-shaped burrows in soft mud. Atlantic, North Sea and Mediterranean. Not uncommon. (**VIII, 23**)

Eupagurus prideuxi (Leach). Up to 4 cm long. Pincers on 1st pair of walking legs are bristled and granular in appearance, right larger than left. Abdomen unsegmented and soft-skinned. Brownish-red. From 10 m downwards, on mud or sand, inhabiting a gastropod shell and usually in association with the anemone *Adamsia*

palliata. Atlantic, English Channel, North Sea and Mediterranean. Common. (**VII, 13**)

Macropodia rostrata (Linnaeus): Long-legged spider crab. Triangular-shaped carapace up to 1·7 cm across. 8 definite spines on carapace. Eyes cannot be withdrawn. Very long 'spidery' legs. Reddish or yellow-brown. In shallow water among seaweeds. Atlantic, English Channel, North Sea, western Baltic and Mediterranean. Common. (**VIII, 1**)

Maia squinado (Herbst): Spiny spider crab. Triangular carapace up to 18 cm long. 2 large points between the eyes and 6 on either side of the carapace, which is covered with small blunt spines on the dorsal surface. Long slender walking legs with small pincers on 1st pair. Yellowish-pink or reddish-brown. Among rocks from shallow water to 50 m. Atlantic and English Channel. Common. (**VIII, 3**)

Cancer pagurus Linnaeus. Edible crab. Carapace up to 30 cm across but often less. Oval in outline with lobed edges resembling a pie-crust; large pincers on 1st pair of walking legs. Pinkish-brown. From shallow water to 100 m, among rocks but sometimes buried in mud or sand. Atlantic, English Channel, North Sea and Mediterranean. Common. (**VIII, 29**)

Phylum: Polyzoa

Pentapora foliacea (Ellis & Solander): Rose coral. Large, conspicuous brittle colonies up to 20 cm high; dome-shaped or spreading on the seabed. Reddish-brown. 15–30 m, on rocky bottoms. Atlantic and English Channel. Common.

Phylum: Echinodermata (Echinoderms)
Class: Crinoidea (Feather stars)

Antedon bifida (Pennant): Rosy feather star. Small concave disc with 5 feathery arms up to 100 cm long. Beneath the disc short processes provide temporary attachment to the substratum. Red, pink, orange, yellow or mauve, sometimes with white markings. On rocks from shallow water to 200 m. Atlantic, English Channel and North Sea. Common. (VIII, 27)

Class: Asteroidea (Starfish)

Porania pulvillus (O. F. Müller). Diameter up to 10 cm, cushion-like with short, stubby arms; tube feet with suckers. Red or orange with white markings. 10–250 m, on gravel. Atlantic, English Channel and northern North Sea. Common. (VIII, 25)

Henricia sanguinolenta O. F. Müller. Diameter up to 20 cm; rounded stiff arms and a small central disc; tube feet with suckers. Purple-orange-red. From shallow water to medium depths on gravel and rocks. Atlantic, English Channel, North Sea and western Baltic. Common. (VIII, 28)

Crossaster papposus (Linnaeus): Common sun-star. Diameter up to 25 cm but usually less. Large disc with 8–15 short tapering arms; tube feet with suckers. Brownish-red with white markings. 10–40 m on sand and gravel and on oyster beds. Atlantic, English Channel, North Sea and western Baltic. Common. (VII, 28)

Marthasterias glacialis (Linnaeus): Spiny starfish. Diameter usually about 30 cm but up to 80 cm. Comparatively small disc and 5 arms, rather soft and covered with spines. Often green but sometimes reddish. Shallow water to 180 m, among rocks or on gravel. Atlantic, western English Channel, North Sea and Mediterranean. Common. (VII, 11)

Class: Ophiuroidea (Brittle stars)

Ophiocomina nigra (Abildgaard). Diameter of disc up to 3 cm, arms about 5 times as long, tapering and covered with long glassy spines. Black, brown or grey. From lower shore to 400 m among rocks or sand. Atlantic, English Channel, North Sea and Mediterranean. Common. (VIII, 30)

Amphiura filiformis (O. F. Müller). Disc diameter up to 1 cm with arms up to 8 cm.

Disc concave between the arms which are very thin and finely spined. Reddish. Shallow water to 100 m, on mud. Atlantic, English Channel, North Sea and Mediterranean. Common.

Class: Echinoidea

Echinus esculentus Linnaeus: Edible sea-urchin. Diameter of test up to 160 mm. Short, dense strong spines covering the almost spherical body. Spines whitish, often with purple tips; test red or purple. From shallow water to 50 m, on rocks or muddy gravel. Atlantic, western English Channel, North Sea. Common. (VII, 12)

Paracentrotus lividus (Lamarck). Diameter of test up to 6 cm; spines up to 3 cm long. Test flattened and spines tapering with sharp points. Small oral opening. Spines purple, brown or greenish; test green. From middle shore in pools, to 30 m. Bores holes in rocks which may be partially enclosed by *Lithophyllum* sp. preventing the exit of the urchin. Often gregarious. Atlantic to the Channel Islands and western Ireland, western English Channel and Mediterranean. Common locally. (VIII, 31)

Class: Holothuroidea (Sea-cucumbers)

Cucumaria elongata Duben & Koren. Length up to 15 cm. Body curved, pointed at hind end and possessing 5 double rows of evenly distributed tube feet. Dark brown. 5–150 m, on mud. Atlantic, English Channel, North Sea and Mediterranean. Common. (VII, 27)

Phylum: Chordata (Chordates)
Class: Ascidiacea (Sea-squirts)

Clavelina lepadiformis (O. F. Müller). Colonial with wedge-shaped individuals up to 3 cm long. Siphons close together. Colonies densely packed but members are only joined by thin stolons at the base. Transparent with red and yellow markings. Shallow water to 50 m, on rocks.

Atlantic, English Channel, North Sea and Mediterranean. Common. (**VII, 21**)

Sub-phylum: Vertebrata (Vertebrates)
Class: Chondrichthyes (Sharks and rays)

Scyliorhinus canicula (Linnaeus): Lesser spotted dogfish. Length up to 70 cm. Blunt snout; nostrils connected to the mouth on the underside by 2 fairly straight grooves. Yellow-grey with small brown spots; under side yellow-white. 6–100 m, among rocks or on sandy bottoms. Egg cases are brown, elongated, horny capsules often found attached by threads to stones or algae. Atlantic, English Channel, North Sea and Mediterranean. Common. (**VII, 7**)

Scyliorhinus stellaris (Linnaeus): Large spotted dogfish or nursehound. Up to 1·5 m long. Similiar to the lesser spotted dogfish but is distinguished by the nasal grooves which do not extend to the mouth, and the 1st dorsal fin which is set well forward above the base of the pelvic fins. Sandy or grey-brown back and sides; under side creamy white. On rocky ground usually in sheltered areas. Atlantic, English Channel, North Sea and Mediterranean. Not particularly common.

Raja clavata Linnaeus: Thornback ray. Length up to 120 cm. Pectoral fins form edge of body giving overall diamond shape. Tail and dorsal surface possess strong spines. Greyish-yellow with dark or light spots; under side white with black around edges. From shallow water to 100 m, on sand or gravel. The egg capsules of this species, sometimes known as mermaid's purse, are shown in the illustration of the sandy shore (**III, 8**). Atlantic, English Channel, North Sea, western Baltic and Mediterranean. Common. (**VII, 22**)

Class: Osteichthyes (Bony fish)

Trisopterus luscus (Linnaeus): Pout or whiting. Length up to 30 cm but often smaller. Deep body; single barbel on lower jaw and characteristic reddish-brown and whitish vertical stripes. From shallow water to 100 m, in shoals, often in caves and wrecks. Atlantic, English Channel, North Sea and western Baltic. Common. (**VII, 8**)

Molva molva (Linnaeus): Ling. Length up to 2 m. Lower jaw shorter than upper and bearing a single barbel. Long tapering body. Adults grey but young have a broad olive-brown stripe along body with vertical black bands. Among rocks from 10–400 m. Atlantic. Common. (**VII, 6**)

Labrus mixtus Linnaeus: Cuckoo wrasse. Length up to 35 cm. Longish body with dorsal fin in a single section. Males are brilliant blue-green on top and light green or yellow on the sides; white on the head indicates breeding condition. Females are red-orange with 3 dark spots on the dorsal surface in front of the tail. Below 10 m, among rocks. Atlantic, English Channel and Mediterranean. Fairly common. (**VII, 2, 3**)

Labrus bergylta Ascanius: Ballan wrasse. Length up to 40 cm. Stocky fish with humped forehead, thick lips and large scales. Dorsal fin lobed at posterior end. Green or reddish-brown; clearly spotted on dorsal surface, often mottled. From shallow water to 12 m, among kelp and rocks. Atlantic, English Channel, North Sea, western Baltic and western Mediterranean. Very common. (**VII, 1**)

Callionymus lyra Linnaeus: Dragonet. Length up to 25 cm. Long, thin, somewhat compressed body; head with large eyes on dorsal surface. Tall, pointed 1st dorsal fin in the male. Females greyish-yellow with brown spots, males orange-yellow with blue spots and orange, blue-spotted fins. From shallow water to 100 m, on sand and gravel. Atlantic, English Channel, North Sea, western Baltic and Mediterranean. Common. (**VII, 24**)

Ammodytes tobianus Linnaeus: Sand-eel. Long slender body up to 20 cm long; long dorsal fin; shorter anal fin; small pectorals and no pelvic fins. Upper jaw swings forward forming an extendable tube. Back yellow-green, sides yellow and under sides silver. Found over fine sand into which they burrow very quickly; sometimes swim head down. From the middle shore down to 30 m. Atlantic, English Channel, North Sea and Baltic. Very common.

Plate V

PLANTS AND ANIMALS OF MUDDY SHORES

1. *Glaux maritima* : Sea-milkwort or black saltwort
2. *Spergularia media* : Perennial sea-spurrey
3. *Juncus maritimus* : Sea-rush
4. *Aster tripolium* : Sea-aster
5. *Limonium vulgaris* : Sea-lavender
6. *Plantago maritima* : Sea-plantain
7. *Polypogon monspeliensis* : Annual beard grass

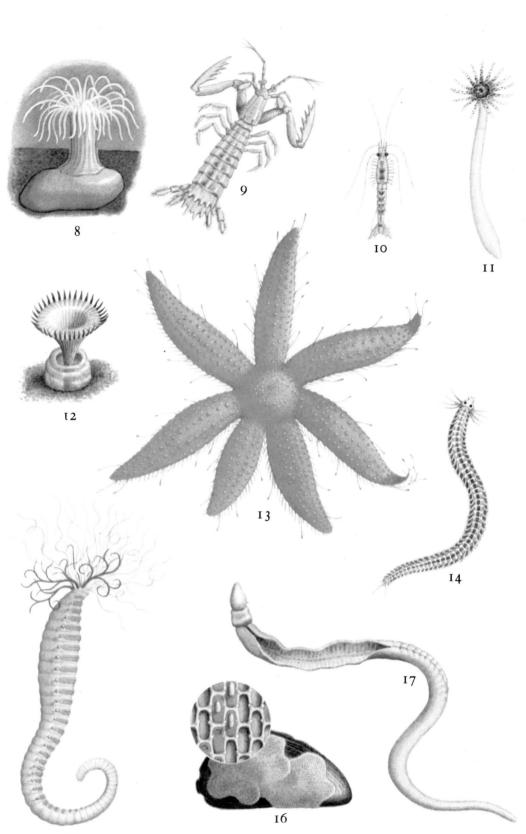

8

9

10

11

12

13

14

15

16

17

Plate V

PLANTS AND ANIMALS OF MUDDY SHORES

8. *Sagartiogeton undata* : Sea-anemone
9. *Squilla desmaresti*
10. *Neomysis integer* : Opossum shrimp
11. *Edwardsia callimorpha* : Burrowing anemone
12. *Myxicola infundibulum* : Tube-dwelling b.istle worm
13. *Luidia ciliaris* : Starfish
14. *Phyllodoce maculata* : Paddle worm
15. *Amphitrite johnstoni* : Terebellid bristle worm
16. *Membranipora membranacea* : Sea-mat, consisting of colonial animals, encrusting a mussel shell. Enlarged detail in circle
17. *Ballanoglossus clavigerus* : Acorn worm

Plate VI

LIFE AROUND THE MUDDY SHORES

1. *Larus ridibundus*: Black-headed gull in winter plumage
2. *Phalacrocorax carbo*: Common cormorant catching a dab
3. *Branta bernicla*: Brent geese feeding on eel-grass
4. *Numenius arquata*: Curlew probing soft mud for hidden animals
5. *Holothuria forskali*: Cotton-spinner or sea-cucumber
6. *Cereus pedunculatus*: Daisy sea-anemone
7. *Zostera marina*: Eel-grass or common grass wrack
8. *Lineus ruber*: Red ribbon worm
9. *Sabella pavonia*: Peacock worm
10. *Hymeniacidon sanguinea*: Sponge
11. *Phallusia mammillata*: Ascidian or sea-squirt
12. *Lutraria lutraria*: Common otter shell
13. *Agonus cataphractus*: Pogge or armed bullhead

Plate VI

LIFE AROUND THE MUDDY SHORES

14. *Spartina x townsendii*: Rice grass or Townsend's cord-grass
15. *Halimione portulacoides*: Sea-purslane
16. *Anser anser*: Greylag geese grazing grassland
17. *Tringa totanus*: Redshank
18. *Salicornia europaea*: Glasswort
19. *Macoma balthica*: Baltic tellin
20. *Scrobicularia plana*: Peppery furrow shell
21. *Aphrodite aculeata*: Sea-mouse
22. *Calidris alpina*: Dunlin in winter plumage
23. *Nereis diversicolor*: Ragworm
24. *Hydrobia ulvae*: Laver spire shell
25. *Turritella communis*: Tower shell
26. *Ocenebra erinacea*: Oyster drill or sting winkle
27. *Prosthecereus vittatus*: Flatworm
28. *Ascidiella aspera*: Sea-squirt
29. *Ostrea edulis*: Common European or flat oyster
30. *Venerupis decussata*: Cross-cut carpet shell or butterfish
31. *Mya arenaria*: Sand gaper or soft-shelled clam

CHAPTER THREE

The Open Sea

The yachtsman making sea crossings or even cruising along the coast spends a considerable amount of time well offshore and often out of sight of land. Sailing on the open sea, he can observe marine life which the land-based naturalist can only find as occasional specimens washed ashore, and he is in an ideal position to make observations which scientists only have the chance to make when aboard their research ships. Because he is less hurried and more in tune with his environment, the yachtsman has a perfect opportunity to discover a world which teems with plankton; he may spot birds which spend much of their time at sea and whales which rarely pass in sight of land. He will occasionally see varieties like turtles which turn up in our latitudes from time to time, and closer inshore he can watch the sea-going side of life of birds like gulls which are land-based but make their living from the sea. The shallow waters of the continental shelf are generally fertile and rich in life but even the deep oceans have much to offer the observant naturalist, especially if he is armed with a plankton net and a microscope.

PLANKTON

The type of plankton caught in a net depends very much on the mesh size used. A large mesh will clearly fail to trap most organisms smaller than the dimension of the mesh—although once the meshes become clogged, a few smaller animals and plants may be caught. Small-mesh nets tend to create more drag when towed through the sea, and so to avoid simply pushing a body of water along in front, a slow towing speed has to be adopted and this will generally allow the larger, faster-moving species to escape the path of the net. (Chapter Five describes plankton nets in some detail.)

The sample collected with a fine net, say with an aperture of 63μ (μ = a thousandth of a millimetre), will be dominated by phytoplankton. The diatoms are usually the most abundant plants of the plankton and include a great range of size and form. Some species like *Thalassiosira* (sp. *excentricus*: **IX, 9**) are large and are always found singly, while many others tend to aggregate and form chains, as *Skeletonema* (sp. *costatum*: **IX, 10**) and *Chaetoceros* (sp. *densum*: **IX, 11**), or, as with *Asterionella* (sp. *japonica*: **IX, 13**), star-shaped colonies. The formation of chains is aided in many instances by the possession of long spines which project from the wall of the cell and link neighbouring cells of the chain together. These spines probably aid, too, in slowing the sinking rate of the cells by increasing the ratio of surface area to cell volume and so increasing the relative amount of frictional drag with the water. There is clearly an adaptive advantage to the plants in sinking slowly, in that they remain as long as possible near the surface where the level of light, necessary for photosynthesis, is comparatively high. The dinoflagellates, with their tail-like flagella, are able to swim weakly and in still water these plants are able to maintain themselves at a level of optimum light intensity. All the plants of the plankton, however, are subject to gross water movements and where vertical mixing occurs due to wave action and circulation, they may be carried insensibly in and out of the photic zone. When conditions are calm and air temperatures are high, thermal stratification stabilises the water column and, provided sufficient plant nutrients are available, will result in a phytoplankton bloom.

Under certain conditions, often associated with calm sunny weather, blooms of flagellates which are able to produce dangerous toxins may appear. These blooms are sometimes termed 'red tides' because they may be very dense and so impart a reddish-brown coloration to the water. Several species may be responsible, for example *Goniaulax tamarensis*, which is illustrated in Fig. 6. Red tides can kill fish and other marine life including birds. Shellfish such as oysters and clams are able to concentrate the toxins in large amounts through their filter-feeding activities, and may give people who eat them severe stomach upsets—some species have even been known to cause death. It is possible that many of the allergies that people reportedly have to shellfish may, in fact, be caused by red tide organisms and not by the shellfish themselves.

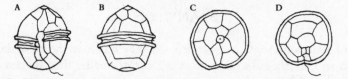

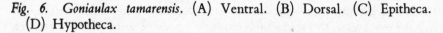

Fig. 6. Goniaulax tamarensis. (A) Ventral. (B) Dorsal. (C) Epitheca. (D) Hypotheca.

Very little is known about red tides and yachtsmen could increase knowledge about them by collecting samples when patches of discoloured water are encountered offshore. Discoloured water inshore is only of interest if it is associated with mortalities of fish or other marine life. Samples should be collected by bucket and stored in a vacuum flask, or refrigerator if possible, and notes should be taken as to time and place, extent of the discoloration, tide, wind, sea, weather conditions at the time, whether there was evidence of fish kills or phosphorescence at night, and any other biological observations.

Medium- and coarse-net plankton samples will contain zooplankton and the kind of animals present will depend on a number of factors. The most important of these will be the geographical location, distance offshore and the time of the year. The zooplankton can be divided into the holoplankton, which includes those species that spend their entire life history in the planktonic phase, and the meroplankton, which spends part of its life history in a benthic or bottom-living phase.

Inshore waters which one can define as stretching as far as the edge of the continental shelf (usually defined as the 200-metre line), are generally characterised by the presence of large numbers of members of the meroplankton—the larval stages of benthic invertebrates and fish which live on or near the seabed in shallow water. Thus we find the larvae of bivalve molluscs, oysters and mussels, etc, with their miniature shells and the ciliated swimming organ, or velum, which is later lost when the larva settles and commences its benthic existence. Gastropods, the snails like winkles and topshells, also have larvae, which possess a tiny whorled shell and a velum drawn out into characteristic rounded lobes. Crustacean larvae are common and some, like the barnacles, may be present in huge numbers. The first stage in the development of the barnacle after hatching is the nauplius with its triangular-shaped carapace and spiny legs. After several moults this changes into an oval-shaped cypris which has a bivalved shell and is the stage which eventually settles, cements itself to the substratum, and changes into the adult barnacle.

The crabs such as *Macropipus puber* (**X, 21**), *Carcinus maenas* (**X, 17**) and *Cancer pagurus* (**X, 16**) have very characteristic larvae known as zoea. These are laterally flattened and they possess a carapace drawn out into two very long curved spines, one on the dorsal surface and the other beneath. These spines probably serve in much the same way as those in phytoplankton cells in slowing the sinking rate of the animal; the zoea swim by means of their much enlarged mouthparts known as the maxillipeds, and the walking legs are hidden beneath the carapace. After two, four or five zoeal stages the crab larva changes into a megalopa; this stage is flattened dorso-ventrally and is rather more crab-like; it swims by means of the abdominal appendages or pleopods and now uses the mouthparts for feeding. The megalopa eventually migrates to the seabed where it becomes transformed into a young crab by a single moult.

Other benthic crustacea: shrimps, porcelain crabs, hermit crabs, Norway lobsters and spiny lobsters, all produce larval stages which exhibit a wide range of shapes and sizes; many of these may appear quite bizarre but all, we may be sure, have evolved their present structure to suit the very special conditions created by the environment and the ecological niche which they occupy. The reason for their appearance may, however, be far from obvious. Why, for instance, does the zoea of the hairy porcelain crab need such an enormous and apparently cumbersome rostral spine? And what is the advantage for the phyllosoma larva of the spiny lobster in being so excessively flattened? Again, they are probably adaptations to slow down the sinking rate. In some species like the lobster the length of time which the larvae spend in the planktonic phase has been reduced to a minimum; the larvae hatch from the eggs, which the mother broods under her abdomen, in an advanced stage of development and are only planktonic for a brief period before they settle to the bottom as miniature lobsters. Larval stages suffer extremely high mortality rates and so there is perhaps an advantage in producing fewer but more robust larvae. This must be at the expense of having to care for the eggs longer while they develop, causing a cessation of growth in the mother as she is unable to moult and increase in size while incubating.

After hatching from the egg, the larvae of the echinoderms—starfish, brittle stars, sea-urchins and sea-cucumbers—are soft-bodied and have a band of cilia running along the body; this is used for swimming. A separate band near the mouth when this develops is for collecting microplankton for feeding. As development continues, the form of the body changes: projections bearing bands of cilia are thrown out and if the larva is of a starfish, a bipinnaria develops; if it is of a brittle star an ophiopluteus; an echinopluteus if it is of a sea-urchin or an auricularia if it is of a sea-cucumber. These are essentially similar but differ in the arrangement of the ciliated bands, and the projections on which they are carried (see Fig. 7). As development continues further, the adult form starts to grow out of the side of the larva, almost like a parasite. Gradually the original larva becomes depleted of its energy reserves until the adult form is ready for settlement, at which point it leaves the plankton and takes up residence on the seabed.

Fig. 7.　Comparison of echinoderm larvae.

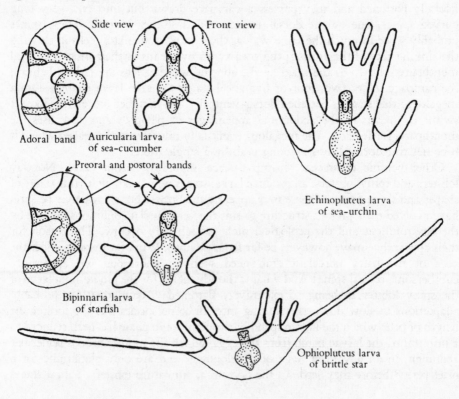

Fish larvae are characteristic of our coastal seas, although they may also be found offshore. The sharks and rays differ from the bony fish in that the eggs are fertilized internally in the female. Most bony fish, however, shed eggs into the water where they are fertilised by sperm from the males. The eggs of some species, like the herring, are shed on the seabed, where they remain attached until hatching, while others commence their planktonic existence immediately. The eggs and larvae of the commercially important fish may be very abundant at certain times of the year in places like the North Sea and English Channel.

The greatest advantage conferred on benthic invertebrates that produce planktonic larval forms is probably distribution of the species, although genetic factors may also be important; the offspring from different populations becomes mixed and so the danger of inbreeding is avoided. Larvae usually live near the surface; they are attracted to light and tend to swim against the force of gravity. They are, therefore, in the best position to be carried far afield by ocean currents. As they approach metamorphosis, when they become bottom-living, they change their response to light and gravity and swim downwards. On reaching the bottom, the substratum is tested and if it is found to be unsuitable, the larva will alight and wait until carried further on, perhaps to a more suitable place. As time passes, most larvae will gradually become less selective of substratum. The type of bottom preferred by different organisms is obviously very variable. Certain worms are attracted to sediments in which members of their own species are already living. Scallops like to settle on seaweed before taking up residence on the seabed. Oysters much prefer clean shell or gravel to settle on; while barnacles require a rocky surface that is within a well-defined vertical range on the seashore.

The pelagic larvae of fish may be carried for long distances with the ocean currents. Some fishes tend to move against the current to their spawning grounds and with the current after they have spawned—on hatching, the larvae are thus carried towards the adult feeding grounds. This behaviour may be seen in plaice off the east coast of Scotland which move north to spawn, swimming against the southerly current; further north, in the waters of the Shetland Islands where the current runs clockwise around the islands, plaice swim anti-clockwise against the current prior to spawning. North Sea plaice migrate to the eastern end of the English Channel in the spawning season and return to the North Sea afterwards; the larvae are later carried back to the North Sea by easterly currents and when they are about 2 cm long they settle to the bottom, close inshore, on the nursery grounds off Denmark. In the North Atlantic, European eels migrate from the rivers of north-west Europe to the Sargasso Sea where they spawn, as mentioned in Chapter One, and then the leptocephalus larvae are carried by the North Atlantic Drift back to the shores of Europe.

There are a great many members of the holoplankton (animals which spend their entire life history in the planktonic phase) in our inshore waters. There is a variety of large jellyfish: the compass jellyfish (*Chrysaora hysoscella*) (**IX, 6**), *Cyanea lamarckii* (**IX, 5**), *Rhizostoma pulmo* (**IX, 8**) and the common jellyfish

(*Aurelia aurita*) (**IX, 7**). The last species in particular may be very abundant at times in estuaries. Swarms of the sea-gooseberry (*Pleurobrachia pileus*) (**IX, 18**), which is a ctenophore, may also appear in huge numbers in estuaries and in offshore waters at certain times.

Of all the groups of animals found in the zooplankton, the crustacea easily predominate, both in total numbers and in the variety of species. Although in freshwater lakes the zooplankton may be dominated by the cladocerans, such as the well-known *Daphnia*, in the sea there are very few members of this group, although some, like *Evadne nordmanni* (**X, 5**) may be common at times.

Certainly the most important crustaceans in the north-west European seas, and in many other sea areas, are the copepods. These animals often dominate plankton hauls. They are generally oval in outline, tending to taper towards the tail. There are two very long antennae which are generally held out to the side, although when swimming with the thoracic limbs, these antennae are held into the side of the body for streamlining. There is one eye in the centre of the head. Although the adult form is easily recognised, there may be juvenile forms present in plankton samples, sometimes in huge numbers. The eggs of most species are carried by the female either in a cluster or in a pair of egg-sacs which are hung beneath the abdomen. The first stage to hatch from the egg is the nauplius, which is the basic type of crustacean larvae also seen in the barnacles and many other groups. The nauplii are kite-shaped. As they grow the cuticle is shed, as with other crustaceans, and with each shedding an increase in size takes place. The nauplius gives rise to a metanauplius and eventually the first copepodite stage which is similar to the adult but with less abdominal segments. There are usually six copepodite stages culminating in the adult.

One of the most important copepods in north-west European waters is *Calanus finmarchicus* (**X, 4**); it is comparatively large and occurs in huge numbers. It is a principle food of the herring and is particularly abundant in the North Sea and around Scotland; further south the very similar species *C. helgolandicus* becomes more important. There are a number of other very common copepods of coastal waters. *Temora* (sp. *longicornis*: **X, 10**) is often abundant in the southern North Sea, and members of the genera *Acartia* (sp. *longiremis*: **X, 6**) and *Centropages* (sp. *typicus*: **X, 8**) are found in the English Channel and southern North Sea, and are often present in estuaries.

The mysids (or opossum shrimps as they are sometimes known, for their habit of carrying the young in a pouch) are characteristic of estuaries. The group includes species such as *Leptomysis gracilis* (**X, 7**), *Mesopedopsis slabberi* (**X, 11**) and *Praunus flexuosus* (**X, 13**). They are very delicate, shrimp-like crustaceans with the eyes carried on stalks. Strictly speaking the mysids are not entirely planktonic, for during the day they are often to be found on the bottom and may be seen among weed and in pools. At night, however, they leave the bottom and swim in the water column.

The oceans beyond the edge of the continental shelf generally have a planktonic fauna distinct from that of coastal waters. It is composed largely of holoplanktonic forms and in general these are widely distributed. Occasionally large forms like the siphonophores, such as the Portuguese man-o'-war (*Physalia physalis*) (**IX, 3**), the by-the-wind-sailor (*Velella velella*) (**IX, 4**) and the helmet-shaped *Muggiaea atlantica* (**IX, 1**) are blown onshore by persistent strong winds and appear in coastal waters or washed up on the beach. The by-the-wind-sailor and the Portuguese man-o'-war both have 'sails' which stand erect above the surface of the sea and enable them to be blown along with the prevailing wind. The by-the-wind-sailor is preyed upon by the beautiful pelagic snail *Janthina* (sp. *janthina* : **IX, 22**) which also floats on the surface by means of a raft of trapped air bubbles. These snails eat all the soft parts of the animal, leaving the horny float and the sail.

The crustaceans in the open ocean are as important as in inshore waters. The euphausids like *Meganyctiphanes norvegica* (**X, 12**) are characteristic of deep seas. They are large shrimp-like animals which resemble the mysids except that they are larger and the thoracic segments are fused with the carapace. The euphausids may occur in great abundance and in the Southern Ocean the species *Euphausia superba* is present in such large quantities that it is the major source of food for the enormous baleen whales. A few amphipod crustaceans like *Parathemisto* (sp. *compressa* : **X, 14**) are present in the plankton but are rarely abundant. The large copepod *Euchaeta norvegica* (**X, 9**) is characteristic of deep waters although it is often found close to shore and in the Norwegian fjords or in deep Scottish lochs. In connection with the crustaceans mention must finally be made of the goose barnacles (sp. *Lepas anatifera* : **X, 2**); these may appear on the hulls of yachts which have been at sea for extended periods and are commonly found attached to driftwood and any other buoyant objects such as the floats of dead by-the-wind-sailors. They are often found washed up on Atlantic shores attached to floating timber.

The urochordates are represented by two classes which commonly occur in offshore waters, although they may often appear closer to the land as well. The first class is the Thaliacea which contains the doliolids and the salps. The doliolids have two generations which alternate between sexual and asexual phases. The sexual phase, or gonozooid, of *Doliolum geganbauri* (**X, 31**) is more common in our waters than the asexual phase or blastozooid. The gonozooid is barrel-shaped, with a mouth at one end and an atrium at the other. It feeds, like its cousins the sea-squirts, by filtering water which is pumped through the body by rhythmic contraction of rings of muscles encircling the barrel. The jet action of the water leaving the atrium also serves to provide the animal with jet propulsion. Eggs are produced which hatch into a tadpole-like larva bearing the urochordate features of a notochord, dorsal nerve chord and pharyngal clefts or gill slits. The free-swimming larva soon loses its tail and becomes barrel-shaped; this is the blastozooid which gives rise to a colony of asexual buds which are carried on a long tail. These split off, develop a further colony and then give rise to the next sexual generation. The salps such as *Salpa democratica* (**X, 32, 33**) also have alternation of generations in their life history but it is much simpler. The sexual form usually occurs in chains and produces an embryo which is released to develop into the solitary, asexual form; this buds off new sexual individuals in chains from a kind of stalk, or stolon, growing on the underside.

The second class of urochordate to be found in the plankton is the group called the Larvacea or Appendicularia. These are pelagic ascidians which retain the larval features right through life. The animal is surrounded by a transparent 'house' through which water is drawn for filtration by the movements of the tail. Larvaceans, like *Oikopleura dioica* (**X, 34**) are readily distinguished from the larvae of benthic ascidians such as *Clavelina lepadiformis* (**X, 37**) by the fact that the tail in ascidians is held in line with the longitudinal axis of the trunk, whereas in the larvaceans the tail is held at right angles to this axis.

As we have seen, different sea areas contain different assemblages of planktonic animals. There are many factors determining whether a particular body of water is suitable for certain species, but the most important of these are likely to be temperature, salinity and the type and amount of suspended matter. Inshore waters generally have a higher percentage of silt particles and organic detritus and also quantities of dissolved organic substances like humic acid which tend to colour the water with a yellowish brown tint. There are some species of zooplankton which have been discovered to be very indicative of the type of water which the sample was taken from. In the waters of north–west Europe two such 'indicator species' have been found to be very useful: *Sagitta elegans* (**X, 23**) is an arrow worm which prefers more oceanic water, while a similar but easily distinguished species *Sagitta setosa* (**X, 26**) is indicative of coastal water. We thus find *S. setosa* in the North Sea and throughout most of the English Channel, while *S. elegans* is found in Atlantic water around the coast of Ireland, in the western approaches to the English Channel and around western and northern Scotland. Further out in the Atlantic yet another species is predominant.

The zooplankton of the Baltic deserves special mention here. The salinity in the Gulfs of Bothnia and Finland is fairly stable and ranges from about 7 ppt at the mouths to about 2 ppt at the inner ends. Areas in the vicinity of river mouths have a variable salinity and it is lowest in spring when the rivers are swollen with melted water from the ice. These two gulfs provide an opportunity for brackish water species to adapt to the low salinity and for freshwater species to adapt to the slightly brackish conditions which exist. At Helsinki in the Gulf of Finland there is a low diversity of species and approximately 60 per cent of those present are of freshwater origin. Moving westwards towards the mouth of the gulf there is a greater diversity of species: the common jellyfish, the cladoceran *Evadne nordmanni*, several copepods such as *Temora longicornis* and mysids like *Praunus flexuosus* and *Neomysis integer* start to appear. The larvae of a number of benthic invertebrates are also likely to be encountered and include those of barnacles and several bivalves: *Macoma balthica*, *Mytilus edulis*, *Cerastoderma edule* and *Mya arenaria*. Similar assemblages of zooplankton are found in the Gulf of Bothnia and outside the two gulfs the diversity is even greater; there is a brackish water form of *Sagitta elegans* in the western Baltic, for instance.

As well as spatial variations in the plankton, there are also seasonal changes. As described in the Introduction, the changes in nutrients brings about blooms of phytoplankton, usually in the spring and late summer. These blooms are generally followed by outbursts of zooplankton population growth as the grazing species are supplied with an abundance of food. The benthic organisms, too, are seasonal in their spawning behaviour—many species spawn in spring, and echinoderm, polychaete and barnacle larvae are common at that time. As a general rule, crustacean and mollusc larvae are present during the summer and some molluscs may also be present in autumn. The zooplankton generally becomes impoverished during the winter months.

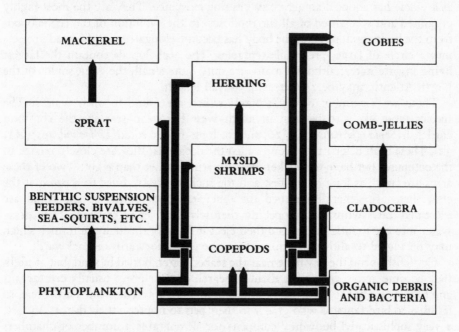

Fig. 8. Simplified food web of zooplankton in coastal seas.

The plankton community provides a very good example of a classic food web whereby several food chains occur alongside each other with many cross-linkages in between. The plankton is of major importance in the economy of the oceans' ecosystems and many interactions occur between the plankton, the benthos, fish, whales, seabirds and, of course, man. Fig. 8 is an oversimplification of a plankton-based food web, but it hints at the complex nature of the system and shows how the various groups of plankton are interdependent. In connection with the question of feeding, it is also worth mentioning the phenomenon of vertical migration. Many species of zooplankton swim upwards during the night and downwards, away from the light, during the day. Not all species carry out these migrations over the same depth range. The outcome of this is that the vertical ranges of many species overlap, which may explain how species in deep water obtain their food supply even though they are below the level of primary production by phytoplankton.

CEPHALOPODS

The class of molluscs known as the cephalopods includes the squid, cuttlefish and octopuses in our waters. The octopus which was mentioned in Chapter Two has secondarily assumed a relatively inactive bottom-dwelling existence, but the class as a whole has adopted an active swimming existence. They are the most highly organised and specialised of all the molluscs. In the evolution of the cephalopods from the basic mollusc plan, the body has become elongated and the head projects into a circle of large, prehensile tentacles. The cephalopods contain the largest living invertebrates: although many are only quite small, the giant squids of the North Atlantic may be 18 metres in overall length.

There are a number of cephalopods which are found in open waters. The species most likely to be seen in north-west European seas are the common cuttlefish (*Sepia officinalis*) (**XI, 26**) and the long-finned squid (*Loligo vulgaris*) (**XI, 32**). These both belong to a group called the decapods; they are closely related to the octopuses but have ten sucker-bearing arms, rather than eight. Two of these arms are twice as long as the rest and the suckers are confined to a patch at the end; these are generally termed the tentacles. Both cuttlefish and squid are extremely fast swimmers, propelling themselves by jet propulsion. They draw water into the mantle cavity and then eject it forcibly through the funnel which may be angled to allow the animal to move either forwards or backwards.

Cuttlefish spend the day time near the seabed, often buried in sand, but at night they become active and swim about in search of their prey; usually crustaceans and fish. At night the cuttlefish becomes more buoyant. The white cuttlebone, so familiar to bird fanciers who give it to their pets to improve their diet, is, in fact, a very sophisticated buoyancy compensator. It contains a number of chambers which at the posterior end are partially gas-filled and partially liquid-filled. At the anterior end they contain gas alone. In order to adjust buoyancy, water is pumped in or out of the chambers by the action of osmotic pressure differences between blood in a membrane which covers the chambers and the liquid inside. The cuttlefish hunts by stalking its prey, or sometimes a jet of water from the funnel may be used to dig crustaceans out of the sand. The tentacles are used to capture the prey, which is killed by biting with the powerful beak. This looks like an upside-down parrot's bill and contains a poison duct which is used to inject the prey.

The long-finned squid is longer and more streamlined than the cuttlefish and is more likely to be found offshore, as it is entirely pelagic. The skeleton has been reduced to a flexible, chitinous 'gladius' or pen and the body is rounded in section. Squid remain in motion continuously and may occasionally be seen leaping from the water at high speed and apparently gliding above the surface for short distances.

FISHES

The sharks and rays are closely related and belong to a primitive class of fish whose skeletons are composed of cartilage and in which there are no bony rays supporting the fins. The tail fin usually consists of two unequal lobes and the vertebral column extends into the larger upper lobe. Although a great deal has been written about sharks, comparatively little is known about their ecology. Probably the best-known members of the class in our waters are the dogfishes. These are sluggish, bottom-living sharks which feed on a variety of crabs, prawns and other benthic invertebrates. The brown, elongated, horny egg-cases of dogfish are often washed up on the beach, and they resemble the egg-case of the thornback ray, sometimes called the mermaid's purse. The lesser spotted dogfish and the rather scarcer larger spotted dogfish or nursehound, described in Chapter Two, may be seen for sale in fishmongers' shops under the exotic name of rock-eel or rock salmon.

There are a number of pelagic sharks which appear in our waters but those most likely to be seen are makos (*Isurus oxyrinchus*) (**XI, 2**), porbeagles or mackerel sharks (*Lamna nasus*) (**XI, 3**), threshers or fox sharks (*Alopias vulpinas*) (**XI, 4**), blues (*Prionace glauca*) (**XI, 5**) and the spur dog or spiny dogfish (*Squalus acanthius*), which is easily identified by having a characteristic spine in front of each dorsal fin. The tope (*Galeorhinus galeus*) (**XI, 6**) is a bottom-living species. Occasional rarities like the hammerhead may turn up from time to time in north-west European waters, but they usually occur further south in the Atlantic. The tope feeds largely on bottom-living species of fish such as members of the cod family and flatfish, but the rest of these sharks feed on open-water shoaling fish like herring, mackerel, scad and pilchards. The thresher shark is said to use its long tail to beat the surface while circling shoals of fish to condense them before going in to the attack.

Special mention must be made of the basking shark (*Cetorhinus maximus*) (**XI, 1**). This is an offshore species, but it may sometimes be seen inshore during the summer, particularly off western coasts of France, Ireland and Scotland. It swims slowly through the water, often with the dorsal and tail fin cutting the surface, and with the mouth wide open. Planktonic organisms are filtered from the water passing through the gill clefts by comb-like gill rakers, which are fine enough to catch the larger members of the zooplankton such as *Calanus*. During the winter months the gill rakers are lost and the shark moves into deeper water where it does not feed. The rakers are then regenerated in time for the next summer's plankton bloom. In the past, basking sharks have been exploited for the sake of their liver oil by local fisheries like that on Achill Island in County Mayo on the west coast of Ireland, but today most of these fisheries have been abandoned. This is perhaps fortunate as it is likely that the slow-breeding basking shark would suffer considerably from overfishing and it would be a great shame if the future of this splendid, peaceful giant were to be endangered.

The rays are close relatives of the sharks, but in the course of their evolution they have become considerably flattened from top to bottom and they live most of the time on the seabed. The mouth and gill openings are on the underside of the body so, when these fish are lying on the bottom, there would be a danger that sand would be drawn in with the stream of water taken through the mouth to oxygenate the gills. There are, however, two openings or spiracles which lie just behind the eyes on the top of the body and these are used to draw in water for the gills when the fish is at rest on the bottom. Once the water has passed over the gills, it is vented through the gill clefts in the normal manner and when the ray is swimming off the bottom, breathing is carried out through the mouth as in other fish.

The thornback, mentioned in the previous chapter, is by far the commonest species of ray in our seas; it feeds on crabs and other shellfish as well as small finfish and it is caught commercially by trawling. The electric ray is an interesting rarity in our waters. There are two species which might be encountered: the common (*Torpedo nobiliana*) and the marbled (*Torpedo marmorata*) (**XI, 10**). They are both rounded in shape, but in the common electric ray the first dorsal fin is larger than the second. Both species are able to deliver a powerful electric shock. The current is produced by two large and two small electric organs which lie on either side of the head. These organs consist of regularly arranged disc-like cells, or electroplates, embedded in a jelly and held in the shape of elongated tubes by connective tissue. The tubes lie vertically in the ray and the plates are arranged so that the upper side of the fish is positive and the lower negative. A peak voltage of 220 volts and a current of 8 amps has been recorded, and it is known that electric shock is used both to stun the ray's prey, which is usually small fish, and to deter predators.

The bony fish of our seas are generally classified according to their habits; demersal fish live on the seabed and pelagic fish live off the bottom in the open sea, often swimming in well-coordinated shoals. As a general rule, pelagic fish tend to have oilier flesh than demersal species, due to their diet of plankton. As has been mentioned, the zooplankton reaches a peak in early summer and autumn and falls off in winter; thus fat reserves have to be built up by the fish during the summer to see it through the lean winter months. Some species, like mackerel, are thought to enter a kind of hibernation during this period.

Typical pelagic fish include the herring, pilchards, mackerel, horse mackerel or scad and the garfish, but others, such as the sunfish and the salmon, also fall into the category. Herring (*Clupea harengus*) (**XI, 8**) and mackerel (*Scomber scombrus*) (**XI, 22**) are two of our best-known commercial fish. They are both caught commercially by drift nets of fine monofilament nylon. The fish swim into the net and are trapped when the gill covers become entangled. Herring are also caught by trawling, and mackerel by purse-seining. In recent years both species have been overfished to a point where the populations are now considerably reduced in many areas.

Herring and mackerel both feed on invertebrate members of the zooplankton, particularly the crustaceans such as copepods and euphausids, which they capture individually. They appear to be deliberate and selective in their feeding and do not just take all planktonic animals indiscriminately. As well as crustaceans, larger fish of both species will also take small fry such as sand-eels, young of various fish including their own species and larger invertebrates—arrow worms, comb-jellies, and so on.

Herring are rarely, if ever, caught on rod and line but, of course, mackerel are frequently taken with a spinner or feathers and they can be very good sport if hooked on light tackle. They are usually caught from middle to late summer when they move inshore and change their diet from plankton to small fish—sprats and sand-eels—which they hunt in small shoals in shallow inshore bays. In late October they then move offshore again and congregate in very large shoals on the bottom, often in hollows and gullies on the seabed where, as mentioned earlier, they are thought to enter into a kind of hibernation. At the end of December these shoals begin feeding near the bottom and start to move towards the spawning grounds in the Celtic Sea, south of Ireland and to the west of the English Channel. They return to the surface and, when they reach their destination, produce their floating eggs over deep water at the edge of the continental shelf. Spawning reaches a maximum in April and the mackerel stay at sea feeding on plankton until midsummer, at which time they start to return to inshore waters again and the shoals divide up into smaller units.

Pilchards (*Sardina pilchardus*) (**XI, 7**) are a member of the same family as the herring. They only occur regularly as far north as the English Channel, where they spawn, but they may reach the North Sea in exceptionally hot summers. The young pilchard is, in fact, the well-known sardine for which there is such an important fishery in Portugal where they are caught for the canning industry by drift or ring nets. Adult pilchards support a small and unimportant fishery in our waters. The horse mackerel or scad (*Trachurus trachurus*) (**XI, 20**), which is not a true member of the mackerel family, and the garfish (*Belone belone*) (**XI, 12**) are of no commercial importance although the garfish is said to be a fine food fish, despite the fact that the bones are turquoise-green in colour.

The Atlantic salmon (*Salmo salar*) (**XI, 11**), which is more usually associated with our freshwater rivers, is an anadromous fish, which means that in common with other species like the sea-trout and the sturgeon (*Acipenser sturio*) (**XI, 9**), it spawns in fresh water, but it may live in the sea for a considerable amount of the time. Some other fish, like the eel, do the reverse and spawn at sea and return to freshwater rivers and lakes as adults. These are termed catadromous. Salmon spawn well upstream, usually in the river in which they were born, and when they hatch the young 'parr', as they are known, may spend from one to four years in the river. At the end of this time they become smolts and migrate to the sea where they live a pelagic life in the northern Atlantic, feeding on herring, mackerel, sand-eels, young cod, euphausids and prawns, and so on. They may spend only one year in the sea, in which case they are called grilse on returning to the river as comparatively small fish, but some may spend up to four years at sea before returning to spawn and it is these fish which attain such large sizes. The adult salmon swim far upstream to where the flow is strong and the river bed is of gravel and the female digs a nest or 'redd', into which she and her partner shed their eggs and milt. These fall into the gravel and the female then digs another nest above the first and the gravel from this is carried down by the current and covers the fertilized eggs. Male salmon parr may join the adults in a redd and contribute towards fertilization, as many of these fish are sexually mature before leaving the river. Once the adult salmon have spawned they lose condition and are known as kelts. These often die, but some manage to return to the sea where they recover and stay for anything up to eighteen months and then return a second time to the river. A very few individuals may repeat the cycle a third or even a fourth time, but this is most unusual.

Salmon farms are becoming an increasingly common sight in the sheltered embayments of the sea coasts of Norway, Scotland and western Ireland. The young salmon are reared in fresh water from artificially spawned eggs and then in order to emulate their natural life cycle they are transferred to floating pens in the sea, where they are fed with dried pellets until they grow to marketable size.

The sunfish (*Mola mola*) (**XI, 30**) is a most extraordinary-looking creature which sometimes appears on the surface near our oceanic coasts in hot calm weather. It swims by undulating the vertical fins from side to side, thus progressing by a kind of swaying motion. Usually, however, these fish are to be seen basking on their side and may disappear below the waves with surprising speed if disturbed. These are essentially oceanic fish which feed on soft-bodied planktonic forms like jellyfish, salps and comb-jellies, but in coastal waters they apparently feed off the bottom as well, because remains of starfish, crustaceans and molluscs have been found in the stomachs of captured specimens.

Turning now to the demersal fish we consider those species which live near the bottom and in general feed on benthic organisms. One of the most important groups of demersal fish, both commercially and to the angler, are the members of the cod family. They are characterised by having soft fin rays and smooth minute scales; most species have a barbel on the chin. The best-known member of the family is the cod itself (*Gadus morhua*) (**XI, 14**). This fish is found from close to the shore to deep water, beyond the edge of the continental shelf. The young cod hatch from pelagic eggs which are spawned over much of the area of the shelf. They at first feed largely on planktonic crustaceans but as they grow they become more dependent on fish, which is almost the exclusive diet of the adult. The pollack (*Pollachius pollachius*) (**XI, 15**) is another very common member of the family. It is an inshore species, usually found in the vicinity of rocks and kelp, feeding on fish, particularly sand-eels, and also bottom-living crabs. Although the pollack may often be caught on spinners or feathers whilst trolling for mackerel, they do not make particularly good eating, as the flesh is rather watery and lacking in flavour. The coley, also called coalfish or saithe (*Pollachius virens*) (**XI, 18**) is similar to the pollack in appearance and is common inshore, particularly in northern waters; it has a greater commercial value than the pollack as it makes better eating.

The flatfishes are well known to most people as they include some of the most flavoursome species to reach the table. They are bottom-living, spending most of the time actually on the seabed, and they are usually superbly camouflaged to match the sand, gravel or mud on which they are resting. Most flatfishes feed entirely on benthic organisms but occasionally they may take off and swim in mid-water and some are active predators of other fish. On hatching from the egg, flatfish larvae are pelagic and have all the appearances of the larvae of any other fish; but then a metamorphosis starts to take place. One eye begins to migrate over the top of the head until it is close to the other one and the whole fish starts to turn over onto what is now its blind side. This side loses all pigmentation and becomes completely white. The pectoral and pelvic fins on this side are smaller, and often the jaw on the lower side is better developed and bears larger teeth than on the upper side. There are three groups of flatfishes in our waters and these have quite distinct characteristics. The first group includes the turbot (*Scophthalmus maximus*) (**XI, 25**) and the brill (*Scophthalmus rhombus*) (**XI, 27**); in these fish the eyes are on the left-hand side of the head, the mouth is at the end of the body and the lower jaw is longer than the upper one. They are less common than the other flatfishes but are particularly good to eat. The second group contains such well-known species as the plaice (*Pleuronectes platessa*) (**XI, 28**) and the flounder (*Platichthys flesus*) (**XI, 29**). The eyes on these fishes are on the right-hand side of the head and again the mouth is at the end of the body and the lower jaw is prominent. The last group is the soles or Dover soles (*Solea solea*) (**XI, 31**) which have eyes on the right-hand side of the head and a small mouth set to one side. The front of the head is smoothly rounded and this feature makes the soles instantly recognisable. They also possess a number of whitish sensory papillae on the underside of the head. All species of flatfish are known to produce reversed specimens in which the eyes are on the opposite side of the head to the usual. This phenomenon is particularly common in the flounder and as the habits and coloration of the fish are turned around as well, it may not be immediately obvious that a fish is reversed.

The bass and the thick-lipped grey mullet are two unrelated fish which are likely to be encountered in shallow water close to the shore. The bass (*Dicentrarchus labrax*) (**XI, 16**) is a great sporting fish and makes excellent eating; it is particularly fond of surf and will come close inshore to feed on small crustaceans which are disturbed from their burrows in the sand by the pounding waves. The thick-lipped grey mullet (*Chelon labrosus*) (**XI, 17**) is a typical algal browser, but larger specimens will also take molluscs and crustaceans. Both bass and mullet are fond of estuaries and brackish waters.

The anglerfish (*Lophius piscatorius*) (**XI, 23**) is a predator which has evolved a most interesting method of catching its prey. It lives most of the time on the seabed, partly concealed by sand and by its very effective camouflage, which includes flaps of skin spread out on the sides of the fish serving to break up its outline. The first fin ray of the dorsal fin has migrated onto the top of the head in the course of its evolution, and on the end of it there is a fleshy growth which is twitched temptingly whenever other small fish are about. Any individual which is drawn towards the lure is then snapped up in a flash by the enormous jaws and swallowed. There are a series of rows of sharp backward pointing teeth arming the jaws and these prevent the escape of larger prey. A great variety of fish have been found in the stomachs of anglerfish and they will also eat crustaceans, lobsters and crabs.

Some interesting adaptations for feeding are also to be seen in some other species of fish. The John Dory (*Zeus faber*) (**XI, 13**) is also known as St Peter's fish because, according to legend, St Peter took a piece of money from its mouth, leaving his thumb print in the form of a dark mark on its side. The John Dory is exceedingly thin and when seen end on is almost invisible; it is thus able to approach its prey unnoticed and at the last moment it strikes by shooting the jaws forwards with great rapidity. The red gurnard (*Aspitrigla cuculus*) (**XI, 24**) has the lowest three rays of the pectoral fin separate from the rest of the fin. These are used for crawling along the seabed and as tactile organs. They are kept in constant movement exploring the bottom; if anything of interest is touched the fish immediately spins round and snaps it up.

TURTLES

Marine turtles are essentially warm water animals; they live in the tropical and sub-tropical regions of the Atlantic and a few live and breed in the Mediterranean. Those specimens reaching north-west European waters are probably strays which have come from the western Atlantic by swimming and drifting from the Gulf Stream and into the North Atlantic Drift.

There are two distinct types of turtle. The Cheloniidae have a shell which consists of large bones covered by horny plates. The commonest member of this group to be seen in our waters is the loggerhead turtle (*Caretta caretta*) (p. 148), but others in the same group include the Kemp's ridley (*Lepidochelys kempii*), the green and the hawksbull turtle (*Chelonia mydas* and *Eretmochelys imbricata*). These latter three are very rare visitors. The second group is the Dermochelydae which contains a single species, the leathery turtle (*Dermochelys coriacea*) (p. 148). The shell in this animal consists of a series of small bones covered by a tough hide-like skin. This is not an unusual visitor to our waters and has even been recorded from the Arctic.

The turtles are excellent swimmers, spending most of their life at sea, and only coming ashore to lay their eggs at the top of sandy beaches. Most species are carnivorous, feeding on soft-bodied planktonic forms, but the green turtle is a herbivore, mostly grazing on eel-grass. Unfortunately, turtles are heavily exploited for their eggs and meat (which is used to make soup) and shell. Suitable breeding areas are also being slowly eroded as isolated sandy beaches are opened up for tourism.

WHALES

Whales, dolphins and porpoises all belong to a group of mammals called the cetaceans. They have evolved from a group of four-legged terrestrial mammals known as the creodonts which took to the sea in comparatively recent geological time. The legs have become considerably modified into flippers, which nevertheless still have the basic bones of hands and feet. A fin on the back and tail flukes have evolved separately, but these do not have a bony skeleton; they are composed of skin and are stiffened by fibrous tissue. The flukes are held horizontally in the whales which swim by vertical undulations of the body; this is in contrast to the fish which swim by lateral movements. The cetaceans have all the characteristics of mammals. They are warm-blooded and have thick layers of blubber to insulate them from the cold waters in which they swim. The young are carried in the body of the mother for a period of months before birth and after they are born they are suckled from teats which lie beneath slits in the mother's skin. This arrangement of the teats maintains streamlining; the male's penis is retracted into a cavity in front of the vent for the same reason. Although the whales have hair like other mammals, this is restricted to a small amount in the region of the chin and snout and is only particularly noticeable in the young. Again, loss of hair is probably an adaptation towards streamlining.

The whales are highly adapted for diving. They are air-breathing and so have to hold their breath when they sound and this restricts the time which they are able to spend under water. Sperm whales may stay down for one to two hours, but this is probably the extreme limit and most whales make shorter excursions to the depths under normal circumstances. The respiratory system of whales is essentially similar to that of other mammals but there are some special features which probably aid them in holding their breath. When we hold our breath, oxygen is depleted in the blood and there is a build-up of carbon dioxide; it is the carbon dioxide which signals the respiratory centre in the brain that it is time to take another breath. In whales the sensitivity of the respiratory centre to carbon dioxide is considerably reduced and so the desire to breathe is delayed. Whales also have high levels of haemoglobin and myoglobin in the blood and muscle. These are the oxygen-carrying and binding pigments in mammals and so increased amounts in the whale allow it to take on more oxygen before diving. When the whale dives, the blood supply to the non-essential, peripheral areas of the body like skin and muscle is cut down and the blood is diverted to the brain and other vital organs which would soon become damaged by lack of oxygen. Whales also have the ability to run up an 'oxygen debt' while they carry out their metabolism in the absence of oxygen; this results in the formation of poisonous lactic acid, which is tolerated until return to the surface where it is eliminated in the presence of oxygen.

When men descend beneath the waves with diving equipment, the nitrogen in the air they breathe becomes dissolved in the blood and body tissues under pressure and, if a diver stays down deep enough and long enough, on returning to the surface the reduction in pressure leads to the formation of bubbles. This happens in much the same way as when bubbles form in beer when pressure in the bottle is released with the top being taken off. In a diver it leads to the condition known as the bends in which bubbles become painfully lodged in the joints or more seriously in the heart, brain or spinal chord. Whales manage to avoid this problem, but again it is not known exactly how. We know, however, that they only take a relatively small amount of air into the lungs when they dive and under pressure much of this is squeezed into the bronchial and tracheal areas where it does not come into contact with blood and so nitrogen uptake is considerably reduced.

The whales may be classified according to whether the teeth have been kept or replaced in the course of evolution by baleen or whalebone. The toothed whales include the common porpoise, the bottlenose dolphin, pilot whale and killer whales. They are largely fish-eaters and so possess teeth to grip their prey, while the baleen whales, like the pike whale or the rare blue whale, feed on plankton. Baleen is the collective term for horny plates which are attached to the roof of the mouth and hang down like curtains. There are fine fringes on the plates and as the whale swims along with its mouth open, plankton such as krill are filtered out by the fringes and the water escapes through the sides of the mouth. From time to time the mouth is closed and the food on the baleen plates is removed by the tongue and swallowed.

In the course of becoming adapted for life in the sea, the toothed whales have developed the ability to produce high-pitched whistles and clicking sounds which are used in the same way as sonar. They receive the echoes produced and these are analysed to locate objects such as prey items, the seabed or other whales. The toothed whales are also able to communicate with each other and it has even been suggested that the extensive range of sounds which they produce represents a simple language. The relatively large brain size of the toothed whales suggests a high degree of intelligence and this is borne out by the fact that dolphins are so easily trained to perform in dolphinariums. The larger baleen whales are less well understood but it seems that they, too, have highly developed acoustic equipment.

How do whales, equipped with such elaborate echo-locating equipment, so frequently become stranded on the shore? It is possible that shallow-sloping shores do not return echoes in the direction of the whale and so unfortunate individuals may carry on swimming towards the shore, unaware of its presence until it is too late. It has also been speculated that ear parasites may sometimes destroy the whale's all-important sense of hearing needed to listen to the echoes of its sonar signals.

In north-west European waters, it is the toothed whales which are most likely to be encountered. The pilot whale or blackfish (*Globicephala melaena*) (**XI, 33**) is common in northern waters but is often seen further south. The common or harbour porpoise (*Phocaena phocaena*) (**XI, 36**) also comes close in shore and up estuaries, and yachtsmen in the English Channel, the North Sea and the Baltic as well as in the open ocean can expect to see the bottle-nosed dolphin (*Tursiops truncatus*). Small schools of these beautiful creatures often come in very close to the shore and will frequently ride the bow wave of a fast-moving boat. The killer whale or grampus (*Orcinus orca*) (**XI, 37**) is usually only seen in the waters of the Atlantic. Although this species has a reputation for ferocity, recent reports suggest that when kept in a dolphinarium they may become quite tame and readily perform tricks like their cousins the dolphins. The whalebone whales are truly oceanic but occasionally sailors may be lucky enough to see one of the commoner species such as the pike whale or lesser rorqual (*Balaenoptera acutorostrata*) closer inland off westerly shores.

BIRDS

The birds which have been described so far have mostly been those of the shore which are probably best seen on excursions away from the boat or while at anchor. We now consider those species which are more likely to be seen at sea. These may be divided into the true oceanic birds which only come ashore occasionally, perhaps only during the nesting season, and those which are characteristic of coastal waters and which roost, nest and sometimes feed ashore.

The auks are probably one of the best-known families of pelagic birds. The family includes the guillemot (*Uria aalge*) (**XII, 12**), razorbill (*Alca torda*) (**XII, 10**) and puffin (*Fratercula arctica*) (**XII, 14**) and also the black guillemot (*Cepphus grylle*) (**XII, 13**) which tends to stay closer to the coast than the other three. All the auks are able to dive to shallow depths and swim under water using their wings, when fishing offshore or searching the bottom for shellfish and worms in shallow water. Far out to sea, one may come across flocks of guillemots, puffins or razorbills flying low over the waves or perhaps resting in 'rafts' on the surface. When they come ashore in the spring to nest, all these birds prefer British and Irish western coasts and they choose isolated rocky cliffs or islands on which to lay their eggs.

Members of the skua family bear a superficial resemblance to the gulls but apart from their dark coloration they may be recognised by their wedge-shaped tails with longer central feathers, and narrow wings which are sharply bent at the wrist. They fly with powerful rapid wing beats and have evolved a rather specialised method of feeding. They exploit the escape reaction of other seabirds, such as the gulls, which will often regurgitate their last meal when attacked in order to hasten their escape. The skuas carry out 'in-flight' attacks, frighten their victims into vomiting and quickly grab the easy meal. A similar strategy is also used by the frigate bird of tropical seas. The skuas can catch fish and will also take carrion or offal from fishing boats. Although they spend most of their time at sea, the skuas come inland to breed on moors and heathland where they build a rough nest of grass and moss. (Great skua—*Stercorarius skua*: **XII, 27**; Arctic skua—*Stercorarius parasiticus*: **XII, 25**.)

The only member of the gull family which can be called truly pelagic is the kittiwake (*Rissa tridactyla*) (**XII, 2**). This bird remains at sea most of the year, feeding on fish, and is commonly seen following trawlers. During the spring kittiwakes come ashore to build quite elaborate nests in colonies on rocky cliffs.

The gannet (*Sula bassana*) (**XII, 17**) belongs to the same group of birds as the pelicans, frigate birds, tropic birds and cormorants, and its closest relatives are the boobies of tropical seas. The gannet soars over the waves on long, pointed wings and executes spectacular dives, sometimes from as high as 30 metres above the waves. It is primarily a fish-eater but will also take waste from trawlers when available. Gannets come ashore to nest on the rocky cliffs of such famous gannetries as Bass Rock in the Firth of Forth, St Kilda in the Outer Hebrides, Mykinesholm in the Faeroe islands and the Saltees on the south coast of Ireland. The nest is usually made of seaweed and the eggs are incubated under the feet.

The last important group of oceanic birds to be seen in the seas off our coasts are the tubenoses, the group that includes the albatrosses, one species of which— *Diomedea melanophris*, the black-browed albatross—may turn up very rarely in the summer months. The more usual representatives in our seas are the fulmars (*Fulmarus glacialis*) (**XII, 7**), shearwaters (sp. *Puffinus puffinus*—Manx shearwater: **XII, 18**) and storm petrels (*Hydrobates pelagicus*) (**XII, 19**). Although very clumsy on dry land, such as when on the nesting ground, these birds are superb fliers and are perfectly adapted for soaring. Whereas terrestrial gliding birds have broad wings, the tubenoses have long narrow ones; the terrestrial gliders and soarers use vertical upcurrents of air such as those in the vicinity of cliffs to provide lift, but the marine equivalent has to use horizontal airstreams and a technique known as dynamic soaring. Birds spending long periods at sea have special problems of salt balance due to the lack of fresh water for drinking and the fact that salt is taken in with the food. The tubenoses, together with other pelagic birds, overcome this problem by secreting very concentrated solutions of salt through a special gland in the nasal cavity. This becomes active when the bird ingests salt and becomes stressed by an excessive salt level in the blood.

Without doubt the gulls are the most conspicuous birds of the coast. They are omnivores and scavengers and great opportunists. The black-headed gull (described in Chapter One) and common gull (*Larus canus*) (**XII, 6**) often feed inland, as well as on the shore, and in estuaries. The herring gull (*Larus argentatus*) (**XII, 4**) is becoming a pest in towns where it is increasingly nesting on buildings. Professor James Green describes in *The Biology of Estuarine Animals* how the black-headed gull tramples around in shallow pools and then steps back to pick up the animals which it has disturbed. This is apparently a good way to collect *Corophium*, and other gulls may use the same trick. The larger species like the herring gull, lesser black-back (*Larus fuscus*) (**XII, 1**) and greater black-back (*Larus marinus*) (**XII, 3**) will steal eggs and young from nests of other birds as well as those of their own species. Crabs and worms are taken from the shore and bivalves may be carried into the air and dropped in order to break the shell. Where the ground is soft mud this is usually ineffective, but gulls in the vicinity of the American hard-shell clam (*Mercenaria*) population of Hamble Spit in Southampton Water have developed the habit of dropping these heavy clams onto the hard shiny roofs of Rolls Royces and Jaguars in the nearby marina car parks. Needless to say this 'interesting behavioural trait' has been greeted with something less than enthusiasm by the owners of those splendid motor cars!

The terns are members of the same family as the gulls. They are easily recognisable, however, being slender with narrow wings, forked tails and sharply pointed bills. Their diet consists almost entirely of small fish such as sand-eels, for which they dive head first into the sea with the wings half folded like the flights of an arrow. Terns nest in colonies which gives them a degree of protection from predators. Any intruding gull or animal is soon spotted and a great clamour is set up; the birds take to the air and with angry screams repeatedly divebomb the threatening interloper until it is forced to retreat. The terns are all summer visitors to European shores; the common (*Sterna hirundo*) (**XII, 9**), roseate (*Sterna dougallii*) (**XII, 8**), sandwich (*Sterna sandvicensis*) (**XII, 11**) and little (*Sterna albifrons*) (**XII, 5**) terns overwinter on the tropical shores of western and southern Africa, while the Arctic tern (*Sterna paradisaea*) spends the northern winter in the summer of the Antarctic, having completed an Arctic to Antarctic migration in as little as six weeks.

The sea ducks are mostly large, heavily built and short in the neck; they dive for their food which consists of molluscs, crustaceans, worms and occasionally a small amount of vegetable material. The common eider (*Somateria mollissima*) (**XII, 23**) breeds around the more northern coasts of Europe, but non-breeding birds may be seen in summer as far south as the English Channel. The scaup (*Aythya marila*) (**XII, 20**), which is the smallest of the ducks described here, arrives in the southern North Sea and English Channel waters in October and feeds close to the shore during the winter, returning in April to its breeding grounds in Iceland and northern Europe. A few may remain to nest in some of the northern Scottish Isles. The common scoter (*Melanitta nigra*) (**XII, 22**) arrives in the southern waters in August/September and although a few non-breeding birds may stay after the following April, the majority return northwards to nest by lakes on the high moors or tundra. The velvet scoter (*Melanitta fusca*) (p. 149) is found on the coasts of northern areas during the winter months, and it moves inland to rivers and lakeshores to nest.

The wigeon (*Anas penelope*) (**XII, 21**) breeds during May in Scotland and northern Europe near rivers or lakes, but during the autumn and winter this surface-feeding duck may be seen on most coastal areas of Europe. Wigeon are highly gregarious and at dusk may be seen moving inland in large numbers to crop the grass in meadows and fields.

The red-breasted merganser (*Mergus serrator*) (**XII, 24**) is not strictly speaking a sea duck, but it is frequently present on saltwater shorelines and in estuaries, particularly during the winter months. It nests in northern England, Scotland, Ireland and the Baltic countries but occurs southwards to a line from northern Spain to Greece in the winter. It is one of a group of fish-eating and diving ducks called sawbills. They have specialised bills with serrations on the sides for gripping prey. They will also eat crabs, shrimps, worms and insects.

There are a few other birds which, although not strictly marine, may be seen by the yachtsman while at sea. The kestrel (*Falco tinnunculus*) (**XII, 26**) can often be seen hovering above a cliff top, sitting motionless on an updraught as it watches for small rodents, insects or small birds which are its staple diet. Or it may be seen equally motionless, except for quivering wing tips, as it faces upwind over sand dunes or saltmarshes. A much rarer sight is the fish-eating osprey (*Pandion haliaetus*) (**XII, 29**). Although this species almost became extinct in Britain and is still critically low in numbers, it is less scarce on mainland Europe and may occasionally be seen hunting for fish in estuaries or the sheltered fjords of Scandinavia. The method of hunting is distinctive: the bird hovers over the prey when it is sighted and then plunges into the water feet first, capturing the fish with its talons.

The chough (*Pyrrhocorax pyrrhocorax*) (**XII, 28**), which is a member of the crow family, may occasionally be seen near rocky cliffs on the coasts of Scotland, the west of Ireland, Wales, Brittany or the Atlantic coast of Spain; and when sailing in the paths of migratory birds, such as the straits of Dover or the Baltic, the yachtsman may spot various migratory birds like starlings, fieldfares or swallows, often in large flocks. Starlings (*Sturnus vulgaris*) (**XII, 30**) resting on lightships in the English Channel sometimes settle in such numbers that every available perching place is taken up.

Birds play an important role in the ecology of marine and coastal ecosystems. As we saw in Chapter One, they are undoubtedly considerable consumers of fish and shellfish populations and many are scavengers in the coastal zone. Birds feeding on fish at sea deposit large quantities of guano on their nesting and roosting grounds and so provide a pathway for organic material to pass from sea to land. There is also a great deal of interaction between human and bird activity. Oil spill disasters, like those of the *Torrey Canyon* in Cornwall and more recently the *Amoco Cadiz* off Brittany, cause huge mortalities of oceanic birds like the auks and others which have to dive for their living and so inevitably become hopelessly fouled in the slick. Increased industry and leisure activity on our coasts seriously reduces the area of suitable feeding and nesting grounds for birds like the little tern, which has become sadly depleted in recent years. Drift nets set for salmon on the west coast of Ireland and off Greenland catch large numbers of auks, particularly razorbills and guillemots, and this could be disastrous in the vicinity of a nesting colony. On the positive side, however, some are able to benefit from man's activities. The fulmar, for instance, has steadily increased in numbers since the turn of the century and this is suggested to be the result of increases in the birds' food resource created by the offal discarded by trawlers. The great black-backed gull, too, seems to be benefiting from the amount of fish offal which is produced by the fishing industry and numbers of this species are also on the increase.

Whether man's activities create an increase or decrease in numbers of a particular species, the important point is that an imbalance has been caused which while benefiting some species will be detrimental to others. Localised increase in the numbers of a voracious, nest-robbing bird, for instance (and the great black-backed gull is a powerful robber), could be disastrous for a nearby nesting colony of another species.

PLANTS AND ANIMALS OF THE OPEN SEA

Plant Kingdom (Phytoplankton)

Division: Algae
Class: Bacillariophycae (Diatoms)

Thalassiosira · excentricus (Ehrenberge) Cleve. Large disc-like cells up to several hundred μ in diameter. Valves sculptured with hexagonal pit-like markings. Many chloroplasts. Neritic. Common. (**IX, 9**)

Skeletonema costatum (Greville) Cleve. Lens-shaped cells; 7–15 μ in diameter. Long parallel spines project from the margins and enable individual cells to unite to form long chains. 1 or 2 chloroplasts per cell. Neritic and estuarine particularly in early spring. Common. (**IX, 10**)

Chaetoceros densum Cleve. Oval cells with flat ends; 20 μ in diameter. At each end of the frustule a pair of long thin spines fuses with neighbouring cells to form long chains. Variable number of chloroplasts. Oceanic. Atlantic, English Channel, North Sea. Common. (**IX, 11**)

Asterionella japonica Castracane. Elongated cells with a thickening at one end; 50 μ in length. The cells stick together at the thicker ends to form star-shaped colonies which may unite to form spirals. Neritic. Common; often abundant in the North Sea and English Channel. (**IX, 13**)

Ditylum brightwellii (West) Grunow ex Van Heurck. Elongated cells 25–60 μ in diameter. At each end there is a pair of spines (or numerous small spines) and a single long central one. Numerous chloroplasts. Neritic. Common; often abundant in warm waters in the Atlantic, English Channel and North Sea. (**IX, 12**)

Biddulphia sinensis Greville. Elongated cells 100 μ in diameter. A pair of spines at each end; valves often triangular when viewed from the end. Single, in twos and threes or in chains. Numerous chloroplasts. First appeared in European waters early 1900s. Neritic and estuarine. Common in English Channel and North Sea. (**IX, 14**)

Rhizoselenia alata Brightwell. Long parallel-sided cells 25 μ in diameter. Short, asymmetric point at each end. Often forming chains. Neritic. Common; often abundant in English Channel and North Sea. (**IX, 15**)

Class: Haptophycae

Phaeocystis pouchetii (Hariot) Lagerheim. Brownish flagellate with a spherical cell embedded in spherical, lobed mucilaginous masses. Often occurs in large numbers. Neritic. Common, particularly in English Channel and North Sea. (**IX, 19**)

Class: Dinophycae (Dinoflagellates)

Protoperidinium depressum (Bailey) Balech. Top-shaped cell with 3 short projecting horns. Cell surface has a transverse and longitudinal furrow and is sculptured to produce hexagonal platelets. Widely distributed. Common. (**IX, 16**)

Ceratium tripos (O. F. Müller) Nitzach. Flattened cell which may be over 200 μ in length. Ornamented theca drawn out into 1 apical and 2 lower horns. Surface has a transverse and longitudinal furrow. Widely distributed. Common. (**IX, 17**)

Animal Kingdom (Zooplankton)

Phylum: Coelenterata (Siphonophores, Hydromedusae and Jellyfish)

Class: Hydrozoa (Siphonophores and Hydromedusae)

Physalia physalis (Linnaeus): Portuguese

man-o'-war. Colony of highly specialised individuals or zooids. Consists of a pneumatophore or gas-filled float up to 30 cm long by 10 cm wide. Below this are suspended the *dactylozooids* which carry stinging cells for hunting prey, the *gastrozooids* for feeding and the *gonodendra* which are the reproductive polyps. Bluish-purple with silver and red on the pneumatophore. Floats on the surface. Dangerous, avoid stinging cells. Atlantic and Mediterranean. May be common after prolonged south-westerly winds. (**IX, .3**)

Velella velella (Linnaeus): By-the-wind-sailor. Round or oval disc up to 8 cm in diameter containing a gas-filled pneumatophore or float and possessing a 'sail' supported by a horny skeleton. Beneath the disc the colony consists of a single *gastrozooid*, for feeding, surrounded by *gonodendra* or reproductive zooids and on the periphery a series of *dactylozooids* for fishing. Bluish-purple with darker fringe; sail transparent and iridescent. Atlantic, English Channel and Mediterranean. Common: occasionally occurs in large numbers. (**IX, 4**)

Muggiaea atlantica Cunningham. Swimming bell is helmet-like and up to 2 cm long. A stem suspended from the bell supports a series of groups of individual polyps each of which includes a feeding zooid, a reproductive zooid and a small swimming bract. Transparent. Atlantic and Mediterranean. Not uncommon. (**IX, 1**)

Sarsia eximia (Allman). Medusa stage of the benthic hydroid *Syncorine eximia*. Large and active; height 4 mm. Stomach contained within margin of umbrella. Transparent. Atlantic, English Channel, North Sea. Common. (**IX, 2**)

Class: Scyphozoa (Jellyfish)

Chrysaora hysoscella (Linnaeus): Compass Jellyfish. Saucer-shaped umbrella up to 40 cm in diameter with 32 lobes around the edge. 24 tentacles, 8 sense organs and mouth arms which extend below the tentacles. Brown spot in centre of umbrella from which radiate 24 brown triangular patches. Atlantic, English Channel and North Sea. Not particularly common. (**IX, 6**)

Cyanea lamarckii. Péron & Leseur. Saucer-shaped umbrella up to 20 cm in diameter with 32 lobes at the periphery. Many tentacles arranged in 8 clusters and 4 frilly mouth arms which are shorter than the tentacles. Bluish-white. Causes severe stings. Atlantic, English Channel and North Sea. Not particularly common. (**IX, 5.**) A similar species, *Cyanea capillata* (Linnaeus), the lion's mane, may be up to 45 cm in diameter. Yellowish-brown. Occurs in the Baltic as well as the Atlantic, English Channel and North Sea. Rare.

Aurelia aurita (Linnaeus): Common jellyfish. Saucer-shaped umbrella up to 25 cm in diameter. 4 mouth arms and numerous small tentacles. 4 purple, crescent-shaped marks when seen from above are the reproductive organs. Colourless and watery; there may be traces of blue. Occurs in estuaries. Atlantic, English Channel, North Sea, Baltic and Mediterranean. Common, may occur in very large numbers. (**IX, 7**)

Rhizostoma pulmo (Macri). Markedly dome-shaped umbrella up to 90 cm in diameter. No tentacles. Periphery divided into 96 small lobes and 16 sense organs. 8 mouth arms descend well below the umbrella and are fused together. Bluish-white or yellowish; mouth arms may have reddish tints. Atlantic, North Sea, western Baltic and Mediterranean. May be abundant locally. (**IX, 8**)

Phylum: Ctenophora (Sea-gooseberries or Comb-jellies)

Pleurobrachia pileus (O. F. Müller): Sea-gooseberry. Gooseberry-shaped body up to 3 cm long. 8 conspicuous rows of fused cilia run from the apex almost to bottom of body. Long tentacles which retract if disturbed. Transparent. Occurs in estuaries. Atlantic, English Channel, North Sea, Baltic and Mediterranean. Common, may be very abundant at certain times. (**IX, 18**)

Phylum : Annelida (Segmented worms)
Class : Polychaeta (Bristle worms)

Tomopteris helgolandica (Greeff). Pelagic worm up to 3 cm long. Paddle-shaped parapodia which end in 2 lobes occur along the length of the body; no bristles. 2 conspicuous palps and 1 pair of eyes on the head. Tentacle-like parapodia behind the head extend approximately $^1/_3$ the body length. Transparent. Atlantic, English Channel and Mediterranean. Common (**IX, 21**)

Phylum : Mollusca (Molluscs)
Class : Gastropoda (Snails)

Janthina janthina (Linnaeus): Violet sea snail. Very thin purple shell 2 cm long with thick terminal portions. Floats upside-down from a raft of bubbles. The eggs are often suspended from an egg raft which is trailed behind. May be found preying on *Velella velella* (by-the-wind sailor). Atlantic. Occasional visitor from sub-tropical waters. (**IX, 22**)

Littorina littorea (Linnaeus): Veliger larva of edible periwinkle. Pale yellow shell up to 600 μ in length with $1^1/_2$ whorls. When active 2 cilated velar lobes with dark pigmented patches are extended from shell. Occurs in estuaries. Atlantic, English Channel, North Sea, western Baltic and Mediterranean. Very common in spring. (**IX, 20**)

Aporrhais pespelicani (Linnaeus): Veliger larva of pelican's foot shell. Yellowish shell has 3 whorls; the last part of the body whorl is striated, the rest are smooth. Characteristic 6-lobed velum up to 2·8 mm across. Each lobe bears a brown spot; whole velum bordered with brown. Atlantic, English Channel, North Sea and Mediterranean. Not particularly common; present in early summer. (**IX, 23**)

Class : Bivalvia (Oysters, Mussels etc.)

Ostrea edulis Linnaeus: Veliger larva of common European or flat oyster. 200–300 μ long with 2 valves. In early veligers the hinge line is straight while later on umbones develop. When active the ciliated velum is extended. Almost transparent. Occurs in estuaries. Atlantic, English Channel, North Sea and Mediterranean. Common. May be very abundant over oyster beds in June and July. N.B. Many other bivalve larvae are similar in appearance and difficult to distinguish.

Mytilus edulis Linnaeus: Veliger larva of the common mussel. Ovoid shell 200–300 μ in length. Well-marked teeth on the hinge line. Transparent. Occurs in estuaries. Atlantic, English Channel, North Sea, Baltic and Mediterranean. Common; occurs mainly in summer and autumn but occasional specimens may be present throughout most of the year. (**X, 1**)

Phylum : Arthropoda (Arthropods)
Class : Crustacea
Subclass : Cirripedia (Barnacles)

Lepas anatifera (Linnaeus): Goose barnacle. Up to 25 cm in overall length; often smaller. The shell, which may be 5 cm long, consists of 5 slightly furrowed plates which are whitish, blue-white or grey sometimes with brown mottling. Stalk up to 20 cm long, slightly retractible, brownish-grey skin. Attached to driftwood, boat hulls, etc., often in large numbers. Atlantic, English Channel and North Sea. Not uncommon. (**X, 2.**) N.B. 3 other similar species occur.

Balanus balanoides (Linnaeus): 5th nauplius and cypris larva of the acorn barnacle. The nauplius has a triangular body up to 900 μ long with a pair of frontal 'horns', a pair of posterior spines and a simple median eye. Semi-transparent. The cypris, which is the stage which settles prior to metamorphosis into the adult form, has a bivalve shell, a compound eye, 6 pairs of thoracic appendages and a short abdomen. Atlantic, English Channel, North Sea, and western Baltic. Very common, particularly in estuaries, March–April. (**X, 3**)

Subclass: Branchiopoda

Evadne nordmanni Loven. Oval-shaped body 900 μ long with small terminal spine. Bivalved carapace with no hinge; head and limbs not covered. No groove between head and body. Large 2nd antenna bearing plumose bristles on the main locomotary appendage. Common, particularly in warmer waters, March–October. (**X, 5**)

Subclass: Copepoda

Calanus finmarchicus (Gunnerus). Elongated oval body 2·5 mm long. 5 distinct thoracic segments, antennules longer than whole body and with a large basal segment. Almost transparent, gut may appear green if full of food; occasional spots of bright scarlet pigment. Particularly abundant in North Sea where it is the principal food of the herring. Uncommon offshore. (**X, 4**.) In more southerly waters it is replaced by *C. helgolandicus* which is almost identical.

Acartia longiremis (Lilljeborg). Small elongated body with 4 obvious thoracic segments. Antennules have a spiky and feathery appearance. Very transparent. In coastal waters and estuaries, very common in northern North Sea. (**X, 6**.) Two other species, *A. bifilosa* (Giesbrecht) and *A. clausi* Giesbrecht, also occur commonly. They are difficult to tell apart but *A. clausi* tends to be a more open ocean species.

Euchaeta norvegica Boeck. Very large body up to 2·5 mm long with 4 thoracic segments and a sharply pointed head. Antennules with very long setae. Rose-red pigment and iridescent 'tail plumes'. Females may carry deep blue eggs.

Particularly common in deep waters of Scottish sea lochs and Norwegian fjords. (**X, 9**)

Temora longicornis (O. F. Müller). Pear-shaped body 1·5 mm long with 4 thoracic segments; appears deep and arched when seen from the side. 2 very long 'tail prongs'. Body has a bluish hue with bright red eye and patches of red and orange pigment. Very common in southern North Sea and eastern English Channel where it replaces *Calanus* as the principal diet of the herring. Common in most coastal waters. (**X, 10**)

Centropages typicus Kroyer. Body 2 mm long with 5 obvious thoracic segments. Head is somewhat square and the characteristic spines on the 5th thoracic segment are asymmetrical. Males possess hinged right antenna and large pincer-like last thoracic legs. Common offshore in English Channel and North Sea. (**X, 8**.) A similar species, *C. hamatus* (Lilljeborg), is more common in coastal waters.

Subclass: Malacostraca

Leptomysis gracilis (G. O. Sars). Slender and shrimp-like with stalked eyes. Up to 1·3 cm long. Pointed rostrum, short carapace, with concave posterior border, leaves last 3 thoracic segments visible from above. Anterior margin of carapace bears a sharp point on either side. Cuticle covered by minute oblong scales. Semi-transparent with reddish-yellow on abdomen. Occurs near the bottom in inshore areas and in pools. May be very abundant in mouths of estuaries at certain times. Atlantic, English Channel, North Sea and Mediterranean. (**X, 7**)

Mesopedopsis slabberi (van Beneden). Long, slender and shrimp-like with exceptionally long eye stalks. Narrow cephalothorax and short carapace which leaves the last 2 thoracic segments exposed. Front edge of carapace rounded. Short telson. Atlantic, English Channel, North Sea and Mediterranean coasts. Common, particularly in brackish waters and estuaries. (**X, 11**)

Praunus flexuosus (O. F. Müller). Long and slender with a narrow carapace which partly covers the last thoracic segment. Up to 2 cm long. Rounded rostrum. Long stout abdomen which curves sharply downwards. In rock pools on lower shore, in estuaries and in shallow water over sand and amongst *Zostera* beds. Atlantic, English Channel, North Sea and Baltic. Very common, often in swarms in estuaries. (**X, 13**)

Meganyctiphanes norvegica (M. Sars): Krill. Carapace, fused with all thoracic segments, constitutes about $^1/_3$ of body length which may be up to 4 cm. Prominent eyes. Short rostrum. Last 7 pairs of thoracic appendages branched near their bases. Open ocean and in deep water. Atlantic, English Channel, North Sea and Mediterranean. Common. (**X, 12**)

Parathemisto gaudichaudi (Guérin). Short body and distinct head with very large eyes. 7–25 mm long. 2 varieties exist: a short-legged form (*P. gaudichaudi* var. *compressa*: **X, 14**) and a long-legged form (*P. gaudichaudi* var. *bispinosa*). Abundant only in colder, northern waters.

Crangon vulgaris Fabricus: 1st larva of the common shrimp. Large zoea 2 mm long, broad telson, large dorsal spine on the posterior margin of the 3rd abdominal segment. In coastal waters and estuaries during spring. Atlantic, English Channel, North Sea, Baltic and Mediterranean. Common. (**X, 19**)

Palinurus vulgaris Latreille: Phyllosoma larva of the spiny lobster. Dorso-ventrally flattened body with rounded outline, very long appendages, transparent. In coastal waters April–June. Atlantic, western English Channel and Mediterranean. Not uncommon locally. (**X, 18**)

Nephrops norvegicus (Linnaeus): 1st zoea of the scampi (Norway lobster or Dublin Bay prawn). Elongated zoea with long spines on the abdominal segments. In

coastal waters during summer. Atlantic, North Sea and Mediterranean. Common locally. (**X, 22**)

Eupagurus bernhardus (Linnaeus): Megalopa larva of the common hermit crab. Length approx. 4·2 mm. 4 pairs of functional pleopods. Conspicuous claws, antennae reach to the tip of the claws. Eye stalks twice as long as they are broad. In coastal waters during summer. Atlantic, English Channel, North Sea, western Baltic and Mediterranean. Common. (**X, 15**)

Porcellana platycheles (Pennant): Zoea of the hairy porcelain crab. Characteristic, long, forward-pointing, rostral spine and 2 fine posterior spines $^1/_2$ the length of the rostral spine. In inshore waters from March to autumn. Atlantic, English Channel, North Sea and Mediterranean. Common. (**X, 20**)

Cancer pagurus Linnaeus: Megalopa of the edible crab. Very pointed rostrum and prominent dorsal spine projecting to the rear and bending upwards near the tip. In coastal waters April–August. Atlantic, English Channel, North Sea and Mediterranean. Common. (**X, 16**)

Carcinus maenas (Linnaeus): 3rd zoea and megalopa of the shore crab. The zoea is approx. 2·8 mm long. Carapace has a long curved rostral and dorsal spine. No lateral spines on the carapace or abdominal segments. Compound eyes with no stalk and forked spiny telson. Distinctive dark chromatophore extending along the side of the carapace. The megalopa is dorso-ventrally flattened and swims using the pleopods on the abdomen. Stalked compound eyes. In coastal waters and estuaries during the summer months. Atlantic, English Channel, North Sea, Baltic and Mediterranean. Common. (**X, 17**)

Macropipus puber (Linnaeus): 4th zoea of the fiddler or velvet swimming crab. Length 4 mm. Long dorsal spine; long straight rostral spine, short lateral spines.

Black and orange markings on the carapace. In coastal waters during early summer. Atlantic, English Channel and North Sea. Common. (**X, 21**)

Phylum: Chaetognatha (Arrow worms)

Sagitta setosa J. Müller. Transparent, narrow, torpedo-shaped body, 2 pairs of lateral fins, the anterior halfway along the body, the posterior terminating just in front of the tail. Mouth possesses strong grasping spines. Transparent. Remains transparent in formalin and is flaccid. Occurs in inshore coastal waters. Atlantic, English Channel, North Sea, western Baltic and Mediterranean. Common. (**X, 26**)

Sagitta elegans Verill. Similar to *S. setosa* but lateral fins more elongated in appearance. Becomes opaque in formalin and is stiff. Indicative of oceanic water. A northern species extending south to the western English Channel Atlantic and North Sea. Common. (**X, 23**.) N.B. *S. elegans* and *S. setosa* rarely occur together in the same sample.

Phylum: Polyzoa

Membranipora membranacea (Linnaeus): Cyphonautes larva of the sea-mat. Triangular body, laterally flattened and possessing a bivalved, calcareous shell. Numerous cilia. Transparent and nearly symmetrical. Oral edge is 0·77 mm long and bears a rounded protuberance. In coastal waters during summer, reaching a maximum in October and November. Atlantic, English Channel, North Sea and Mediterranean. Common. (**X, 30**)

Phylum: Echinodemata (Echinoderms)
Class: Asteriodea (Starfish)

Asterias rubens Linnaeus: Bipinnaria and brachiolaria larva of the common starfish. Bipinnaria (**X, 24**) is about 2 mm long with 12 symmetrical lobes or arms which are extremely elongated in this species. The brachiolaria stage (**X, 25**) possesses 3 special appendages, in the pre-oral region, bearing numerous adhesive papillae. In coastal waters May–September. Atlantic, English Channel, North Sea and western Baltic. Common.

Class: Ophiuroidea (Brittle stars)

Ophiothrix fragilis (Abildgaard): Ophiopluteus larva of the common brittle star. Conical in shape and about 0·3 mm long. Apex of cone is the posterior end and bears 4 pairs of arms. 1 pair of these is characteristically very long in this species. In coastal waters April–August. Atlantic, English Channel, North Sea and Mediterranean. Often very abundant. (**X, 28**)

Class: Echinoidea (Sea-urchins)

Psammechinus miliaris (Gmelin): Echinopluteus larva of the green sea-urchin. Laterally compressed cone-shaped body, the apex of which is prolonged to form an appendage. 4 pairs of arms. 0·5–1 mm long. In coastal waters during the summer. Atlantic, English Channel,. North Sea and western Baltic. Not uncommon. (**X, 27**)

Class: Holothuroidea (Sea-cucumbers)

Holothuria sp. Pentacula larva of a sea-cucumber. Larva resembles a small sea-cucumber with 5 appendages in front of the mouth and a series of ciliated bands running around the body. In coastal water during spring. (**X, 29**)

Phylum: Chordata (Chordates)
Subphylum: Urochordata
Class: Thaliacea (Salps and Doliolids)

Doliolum geganbauri (Uljanin). Transparent, barrel-shaped body, up to 9 mm long, encircled by 8 colourless or blue muscle bands. Late summer. Atlantic Ocean species which strays into the western English Channel and northern

North Sea. Abundant at times. (**X, 31**)

Salpa democratica Forskal. Occurs in 2 forms: solitary (**X, 32**) and aggregated (**X, 33**). Solitary individuals may be up to 1·5 cm long, cylindrical, transparent, with 6 muscle bands—the 2nd 3rd and 4th fused dorsally and the 5th and 6th also fused dorsally. Aggregated individuals are up to 0·6 cm long with 4 muscle bands, the 1st 3 being fused dorsally. The colonies often have long streamers up to 30 cm. Warm water species. Atlantic, English Channel, North Sea and Mediterranean. May be abundant.

Class: Larvacae (Appendicularians)

Oikopleura dioica Fol. Oval-shaped body or trunk and narrow tail tapering to a point. Overall length 6–7 mm, tail length 2–4 mm. In life the animal is surrounded by a fragile, gelatinous 'house' which is usually destroyed in netted plankton samples. In coastal waters and estuaries during spring. Atlantic, English Channel and North Sea. Common. (**X, 34**)

Class: Ascidiacea (Sea-squirts)

Clavelina lepadiformis (O. F. Müller): Larva of sea-squirt. Square-shaped body or trunk with long tapering tail which is held in line with the longitudinal axis of the trunk. Coastal waters. Atlantic, English Channel, North Sea and Mediterranean. Not uncommon. (**X, 37**)

Subphylum: Vertebrata (Vertebrates)

Solea solea (Linnaeus): Embyro and newly hatched larva of the sole. Egg capsule

(**X, 36**) undulated and punctured, ring of characteristic oil droplets arranged on underside of embryo which is whitish-yellow. Early larva (**X, 35**) 3 mm long; yolk sac, marginal fin and body covered with yellowish-white spots. At 5 mm yellow-ochre pigment covers abdomen and dorsal and ventral body margins, black spots present. At 11 mm prior to metamorphosis colour is black and orange with yellow patches. March–June. Atlantic, English Channel, North Sea, western Baltic and Mediterranean. Common.

Clupea harengus Linnaeus: Early and post-larval stages of the herring. Newly hatched larvae (**X, 40**) 5–6 mm long, devoid of coloration except black eyes, elongated. At 10 mm the head is swollen and later an increase in depth and breadth relative to length occurs. Early larvae (**X, 38**) occur near the bottom and migrate to mid-water at the stage when the larval fin is replaced by the dorsal fin. Time of appearance varies: March in the North Sea, December–March in the western English Channel. Atlantic, English Channel, North Sea and Baltic. Common.

Sardina pilchardus (Walbaum): Embryo (**X, 42**), and newly hatched larva (**X, 41**) of the pilchard (or sardine when small). Egg yolk marked by polygonal reticulations. No pigment at all in the early larva which is about 3·8 mm long. As the jaw elongates and mouth develops eyes become yellow and later black. May–late autumn. Atlantic north to Ireland, English Channel and North Sea. Common.

Larger Animals of the Open Sea

Phylum: Mollusca (Molluscs)
Class: Cephalopoda (Squid, Cuttlefish and Octopus)

Loligo vulgaris Lamarck: Long-finned squid. Up to 50 cm long. Elongated, torpedo-shaped body with paired triangular fins running halfway along body and joined at the end. 8 short arms with 2 rows of suckers; 2 long arms with unequal suckers in 4 rows on the ends. Internal horny shell or 'pen'. Pink, white, purple and brown; mottled. Usually offshore. Atlantic, English Channel, North Sea (occasionally), western Baltic (occasionally) and Mediterranean. Not uncommon. (**XI, 32**)

Sepia officinalis (Linnaeus): Common cuttlefish. Up to 30 cm long. Broad, flat, oval-shaped body with lateral fins. Conspicuous funnel on underside near the head. 8 short arms each with 2–4 rows of suckers; 2 long arms with unequal suckers in 4 rows on the ends. Yellow, green, brown and black; striped or mottled; changes colour rapidly when disturbed. Usually over sand. Atlantic, English Channel, North Sea and Mediterranean. Common. (**XI, 26.**) The white 'cuttlebone' is shown on the illustration of the sandy shore (**III, 6**).

Phylum: Chordata (Chordates)
Subphylum: Vertebrata (Vertebrates)
Class: Chondrichthyes (Sharks, Skates and Rays)

Cetorhinus maximus (Gunnerus): Basking shark. Up to 11 m long. Stout body; moon-shaped tail; very long gill slits running from the back behind the head to the throat; very small teeth. Greyish-brown or black dorsally, lighter ventrally. Oceanic but may occur onshore in calm, fine weather during summer. Atlantic, English Channel, northern North Sea, western Baltic and Mediterranean. Common (**XI, 1**)

Isurus oxyrinchus Rafinesque: Mako. Up to 4 m long. Streamlined body with a conspicuous keel on either side of the body in front of the tail; sharply pointed snout; origin of 1st dorsal fin vertically behind pectoral fin; 2nd dorsal fin originates in front of the anal fin. Large sharp teeth. Deep blue-grey, white underneath. Oceanic shark swimming near the surface. Atlantic and western approaches of English Channel. Relatively uncommon. (**XI, 2**)

Lamna nasus (Bonaterre): Porbeagle or mackerel shark. Up to 3 m long. Large thick-set body; smoothly rounded snout and distinct keel on either side of the tail; origin of 1st dorsal fin above base of pectoral fins; 2nd dorsal fin directly above the anal; large triangular teeth. Dark blue or bluish-grey; pale cream on undersides. Oceanic shark usually seen offshore over deep water. Atlantic, English Channel, North Sea and Mediterranean. Not uncommon. (**XI, 3**)

Alopias vulpinus (Bonaterre): Thresher or fox shark. Up to 6 m long; tail more than $^1/_2$ total length; long curved pectoral fin; minute 2nd dorsal and anal fins; short, rounded snout; eyes set well forward; small triangular teeth. Brown, grey, blue or black on the back; white underside. Oceanic shark occasionally coming inshore. Atlantic, English Channel, northern North Sea and Mediterranean. Not particularly common. (**XI, 4**)

Prionace glauca (Linnaeus): Blue shark. Up to 3·8 m long. Long pointed snout and elongated, curved pectoral fins. Triangular teeth. Dark, indigo blue back, brilliant blue sides and brilliant white underneath. Open ocean species; near the surface. Atlantic, western English Channel and Mediterranean. Common. (**XI, 5**)

Squalus acanthia (Linnaeus): Spur dog or spiny dogfish. Up to 1·2 m long. Easily distinguished by spines at anterior end of each dorsal fin. In shallow water and deeper, over a variety of ground. Atlantic, English Channel, North Sea and Mediterranean. Common.

Galeorhinus galeus (Linnaeus): Tope. Up to 1·5 m long. Small shark with pointed head; sharp pointed teeth. Grey or light brown with light-coloured belly. Shallow water, bottom-living shark often found over gravel. Atlantic, English Channel, North Sea and Mediterranean. Common. (**XI, 6**)

Torpedo marmorata Risso: Marbled electric ray. Up to 60 cm long. Disc-shaped body; 2 rounded dorsal fins similar in size. Dark with light mottling on top; cream-coloured beneath. In shallow water usually over sand. Atlantic, north to south-west Britain and Ireland, and Mediterranean. Common in southern waters but rarer north of Brittany. (**XI, 10**)

Torpedo nobiliana Bonaparte: Common electric ray. Up to 1·8 m long. Rounded, disc-like body; 1st dorsal fin distinctly larger than the second; inner edges of the spiracles are smooth. Dorsal surface dark grey or brown; under surface light. Bottom-living usually on mud or sand. Atlantic, English Channel, North Sea and Mediterranean. Not uncommon.

Class: Osteichthyes (Bony fishes)

Acipenser sturio Linnaeus: Sturgeon. Length up to 3·5 m. An unmistakable species. 5 rows of large, hard, bony plates along the length of the body; long snout; 2 pairs of barbels. Bluish-black on the dorsal surface, lighter on the sides and white underneath. A marine fish which enters large rivers, particularly the Gironde and Dordogne in France, in early spring to spawn. Atlantic, English Channel, North Sea, Baltic and Mediterranean. Rare. (**XI, 9**)

Clupea harengus Linnaeus: Herring. Length up to 40 cm. Silvery fish with large, easily detachable scales. Lower jaw prominent and longer than upper jaw. Origin of the dorsal fin in front of, or above, the base of the pelvic fins. A pelagic species generally found offshore. Atlantic, English Channel, North Sea and Baltic. Less common than formerly. (**XI, 8**)

Sardina pilchardus (Walbaum): Pilchard or sardine (when small). Up to 26 cm long. Herring-like rounded body with large scales; gill cover strongly marked with pronounced radiating ridges; jaws equal in length. Origin of dorsal fin in front of the base of the pelvic fin; greenish dorsal surface, sides golden or silvery and white ventrally. A pelagic fish occurring offshore in large shoals. Atlantic north to Ireland, English Channel and occasionally in the North Sea during hot summers. Common. (**XI, 7**)

Salmo salar. Linnaeus: Atlantic salmon. Up to 1·5 m long. Large jaws, the upper extending back to the posterior margin of the eye; in the male the jaw may become hooked prior to breeding. Distinguished from trout by the narrow tail in front of the caudal fin. Marine fish entering rivers to spawn. At sea colour silvery blue or green on the back, silvery sides and white belly; in rivers silvery appearance is lost and fish becomes green or brown with orange or red mottling. Atlantic north of Biscay, English Channel, North Sea and Baltic. Not uncommon. (**XI, 11**)

Belone belone (Linnaeus): Garfish. Up to 76 cm long. Elongated fish with long beak-like jaws heavily armed with sharp teeth; head more than $1/4$ the body length. Brilliant green or dark blue dorsally, sides silver, undersides yellowish. Oceanic but entering coastal waters regularly. Atlantic, English Channel, North Sea, Baltic and Mediterranean. Common. (**XI, 12**)

Pollachius pollachius (Linnaeus): Pollack. Length up to 80 cm. Protruding lower jaw; no barbel; large eye; lateral line arched over pectoral fin; short pelvic fins. Dark brown or green back, silvery-yellow sides, light underneath. Inshore over rocks or swimming in mid-water. Atlantic, English Channel, northern North Sea, western Baltic and Mediterranean. Common. (**XI, 15**)

Gadus morhua Linnaeus: Cod. Length up to 1·5 m. The dorsal and anal fins are rounded; small eye; lateral line curves over pectoral fin and straightens under middle dorsal fin; conspicuous barbel on chin. Sandy brown or greenish dorsal surface and sides with brown markings, light underneath. In shallow water and in deep water offshore. Atlantic, English Channel, North Sea and Baltic. Common. (**XI, 14**)

Plate VII

THE DIVER'S WORLD

1. *Labrus bergylta* : Ballan wrasse
2. *Labrus mixtus* : Cuckoo wrasse (female)
3. *Labrus mixtus* : Cuckoo wrasse (male)
4. *Chondrus crispus* : Carragheen or Irish moss
5. *Caryophyllia smithi* : Devonshire cup coral
6. *Molva molva* : Ling
7. *Scyliorhinus canicula* : Lesser spotted dogfish
8. *Trisopterus luscus* : Whiting or pout
9. *Padina pavonia* : Peacock's tail
10. *Homarus gammarus* : Common lobster
11. *Marthasterias glacialis* : Spiny starfish
12. *Echinus esculentus* : Edible sea-urchin
13. *Eupagurus prideuxi* : Hermit crab wearing cloak anemone (*Adamsia palliata*)

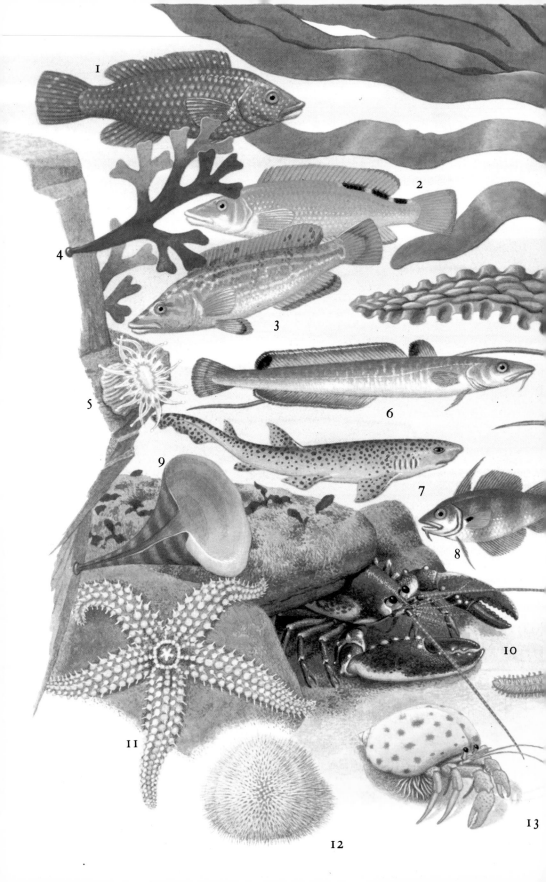

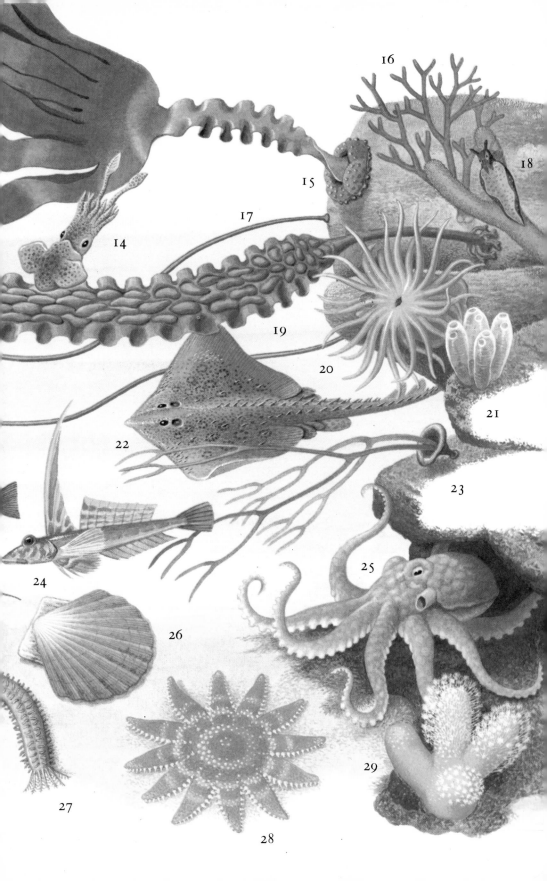

Plate VII

THE DIVER'S WORLD

14. *Sepiola atlantica*: Little cuttlefish
15. *Saccorhiza polyschides*: Furbelows seaweed
16. *Codium tomentosum*: Sponge seaweed
17. *Chorda filum*: Bootlace weed
18. *Elysia viridis*: Sacoglossan sea-slug
19. *Laminaria saccharina*: Sugar kelp or poor man's weather glass
20. *Anemonia sulcata*: Snakelocks anemone
21. *Clavelina lepadiformis*: Colonial ascidian
22. *Raja clavata*: Thornback ray
23. *Himanthalia elongata*: Thong weed
24. *Callionymus lyra*: Dragonet (male)
25. *Eledone cirrhosa*: Lesser or curled octopus
26. *Pecten maximus*: Great scallop
27. *Cucumaria elongata*: Sea-cucumber
28. *Crossaster papposus*: Common sun-star
29. *Alcyonium digitatum*: Dead man's fingers

Plate VIII

PLANTS AND ANIMALS OF THE DIVER'S WORLD

1. *Macropodia rostrata* : Spider crab
2. *Limacia clavigera* : Nudibranch
3. *Maia squinado* : Spiny spider crab
4. *Glycymeris glycymeris* : Dog cockle
5. *Facelina auriculata* : Nudibranch
6. *Aplysia punctata* : Sea-hare
7. *Chlamys varia* : Variegated scallop
8. *Gibbula magus* : Topshell
9. *Clathrus clathrus* : Common wentletrap
10. *Trivia monacha* : European cowrie
11. *Haliotis tuberculata* : Green ormer
12. *Bispira volutacornis* : Tube-dwelling polychaete worm
13. *Chlamys opercularis* : Queen scallop
14. *Pennatula phosphorea* : Phosphorescent sea-pen
15. *Dilsea carnosa* : Red seaweed
16. *Delesseria sanguinea* : Red seaweed
17. *Metridium senile* : Plumose anemone
18. *Actinothoë sphyrodeta* : Sea-anemone
19. *Corynactis viridis* : Jewel anemone
20. *Modiolus modiolus* : Horse mussel
21. *Plumularia catharina* : Sea-fir
22. *Suberites domuncula* : Sea-orange

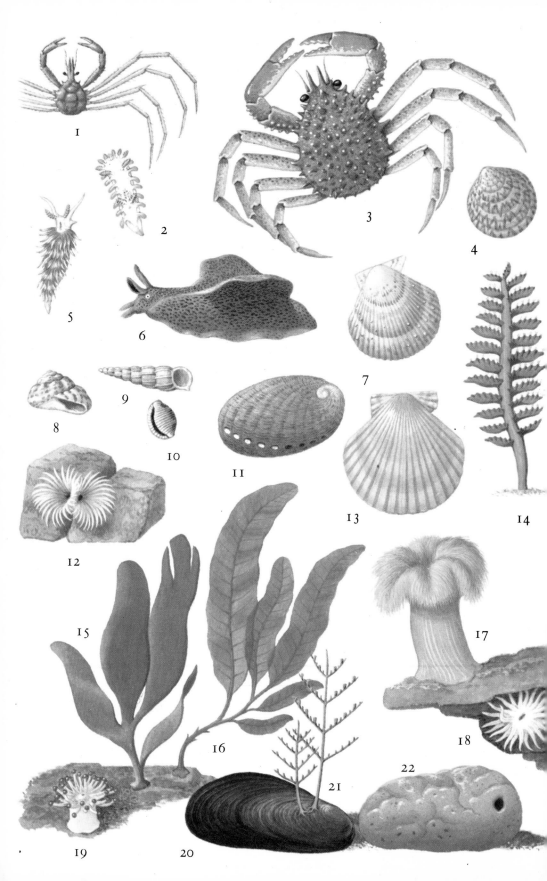

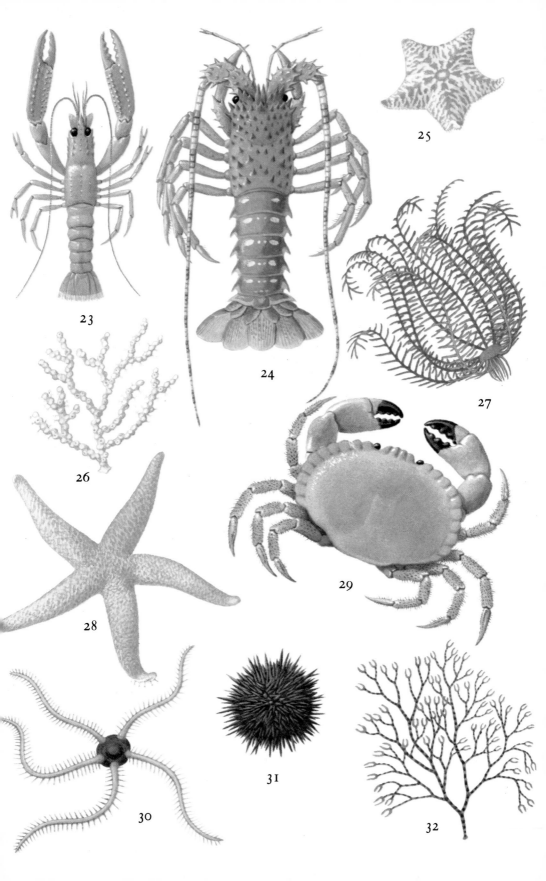

23

24

25

26

27

28

29

30

31

32

Plate VIII

PLANTS AND ANIMALS OF THE DIVER'S WORLD

23. *Nephrops norvegicus*: Norway lobster, Dublin Bay prawn or scampi
24. *Palinurus vulgaris*: Spiny lobster
25. *Porania pulvillus*: Cushion star
26. *Eunicella verrucosa*: Sea-fan
27. *Antedon bifida*: Rosy feather star
28. *Henricia sanguinolenta*: Starfish
29. *Cancer pagurus*: Edible crab
30. *Ophiocomina nigra*: Brittle-star
31. *Paracentrotus lividus*: Sea-urchin
32. *Ceramium rubrum*: Red seaweed

Pollachius virens (Linnaeus): Saithe, coalfish or coley. Length up to 60 cm. Lower jaw projects slightly beyond the upper; very small barbel on the chin; lateral line straight. Greenish-brown on dorsal surface, silvery-grey on sides and ventral surface. Inshore but deeper than the pollack. Atlantic, English Channel, North Sea and western Baltic. Common. (**XI, 18**)

Zeus faber Linnaeus: John Dory. Up to 40 cm long. Deep, flattened head and body; massive jaws; tall dorsal spines; 4 massive spines in advance of the rays on anal fin; long pelvic fins; sharp spines on belly and along sides of dorsal and anal fins. Yellowish-brown or grey, lighter on belly, conspicuous black spot ringed with yellow in middle of each side of the body. Solitary fish occurring inshore and in deeper water. Atlantic, English Channel, North Sea, western Baltic and Mediterranean. Scarce. (**XI, 13**)

Chelon labrosus (Risso): Thick-lipped grey mullet. Up to 60 cm long. Small mouth with very thick upper lip; small teeth; 1st dorsal fin has 4 rays. Dorsal surface dark greenish-grey or blue, sides and belly silvery. Inshore and in estuaries often near marinas. Atlantic, English Channel, southern North Sea and Mediterranean. Very common. (**XI, 17**)

Dicentrarchus labrax (Linnaeus): Bass. Length up to 1 m. Long-bodied with large scales; spiny dorsal fin; teeth on hind edge of front gill cover and 2 spines on main gill cover. Grey or blue-back, silver flanks and yellowish-white belly. Generally inshore among rocks and surf; also in estuaries during summer. Atlantic, English Channel, North Sea, western Baltic and Mediterranean. Common. (**XI, 16**)

Mullus surmuletus Linnaeus: Red mullet. Length up to 30 cm. 2 large barbels; deep curved snout; 2 large scales on cheek below the eye. Characteristic reddish coloration with longitudinal yellow stripes; scales on back edged with brown; dark stripe near top of 1st dorsal fin. Occurs in shallow

water over rough ground or on sand and gravel. Atlantic, western English Channel and Mediterranean. Fairly common. (**XI, 21**)

Trachurus trachurus (Linnaeus): Horse mackerel or scad. Length up to 35 cm. Large heavy head; large jaws with the lower protruding; large bony plates run all along the lateral line and are spiny near the tail; 2 dorsal fins, the 1st with tall rays; 2 spines precede the anal fin. Grey-blue or greenish on dorsal surface; silvery sides; white belly. Swims in shoals in open water. Atlantic, English Channel, North Sea, western Baltic and Mediterranean. Common. (**XI, 20**)

Scomber scombrus Linnaeus: Mackerel. Up to 40 cm long. Rounded, elongated body; 2 dorsal fins and a row of finlets running along the tail behind the dorsal and anal fins; deeply forked tail. Dorsal surface shiny blue; flanks metallic, belly white. Characteristic black curving lines extend down the back. Pelagic; swims near the surface in large shoals inshore and offshore. Atlantic, English Channel, North Sea, western Baltic and Mediterranean. Common. (**XI, 22**)

Aspitrigla cuculus (Linnaeus): Red gurnard. Length up to 30 cm. Concave snout ending in 3 or 4 sharp spines; pectoral fins just reach as far as the vent; lateral line has short soft lateral extensions. Deep red coloration, lighter on belly; pelvic fins rosy, anal fin white at base. From 20 m downwards on sand, gravel and amongst rocks. Atlantic, English Channel, North Sea, western Baltic and Mediterranean. Common except in North Sea. (**XI, 24**)

Cyclopterus lumpus Linnaeus: Lump sucker or sea-hen. Length up to 55 cm. Rounded, thick-set, deep body with short, rounded head; head and body covered with distinct bony 'teeth' and 4 rows of large bony plates. Ventral sucker disc formed from the pelvic fins. Greyish with lighter undersides. In the breeding season males are bluish dorsally and orange underneath.

From shallow water to 300 m. Atlantic, north of Portugal, English Channel, North Sea and Baltic. Not uncommon. (**XI, 19**)

Lophius piscatorius (Linnaeus): Anglerfish. Length up to 2 m. Broad head and very large mouth with large curved teeth in upper and lower jaws; no scales but many bony plates on the skin; a series of long thin rays run along the mid-line of the head. Brown, green or reddish; white underneath. From shallow water to 500 m. Atlantic, English Channel, North Sea and Mediterranean. Common in deeper water; occurs occasionally inshore. (**XI, 23**)

Scophthalmus maximus (Linnaeus): Turbot. Length up to 60 cm. Mouth to the left of the eyes; large strongly curved jaws; broad, rounded body with length $1^1/_2$ times the width. Neither dorsal nor anal fins extend beneath the tail. No scales but bony plates scattered over body and small bony plates on eyed side of head. Dull brown, with brown specks over body and tail fin. On muddy sand and gravel from shallow water to 100 m. Atlantic, English Channel, North Sea, Baltic and Mediterranean. Common. (**XI, 25**)

Scophthalmus rhombus (Linnaeus): Brill. Length up to 50 cm. Mouth to the left of the eyes; large jaws; mouth curved. Body oval and length $1^3/_4$ times the width. Neither dorsal nor anal fin continues under the tail; 1st rays of dorsal fin branched for $1/_2$ their length. Moderately large scales and no bony plates on skin. Sandy-brown or grey with dark blotches and spots. In shallow water and down to 80 m on gravel sand or mud. Atlantic, English Channel, North Sea, western Baltic and Mediterranean. Common. (**XI, 27**)

Platichthys flesus (Linnaeus): Flounder. Up to 20 cm long. Small, terminal mouth usually to the right of the eyes but often to the left. Head more than $1/_4$ of the body length; 1 or 2 bony knobs at the beginning of the lateral line; lateral line curved above the pectoral fin. Small body scales; coarsely toothed scales on either side of lateral line

and characteristic prickles along bases of dorsal and anal fins. Dull brown, grey or green sometimes with pale orange speckles. In shallow water and down to 50 m; occurs in estuaries. Atlantic, English Channel, North Sea, Baltic and Mediterranean. Very common. (**XI, 29**)

Pleuronectes platessa Linnaeus: Plaice. Length up to 55 cm. Mouth terminal and to right of the eyes; a series of 4–7 bony knobs runs back in a curve from the eyes to the start of the lateral line. Upper side brown with distinctive orange spots. On soft bottoms from shallow water down to 350 m. Atlantic, English Channel, North Sea, western Baltic and Mediterranean. Common. (**XI, 28**)

Solea solea (Linnaeus): Sole or Dover sole. Length up to 30 cm. Oval body; blunt head with mouth at front extremity. Pectoral fins on upper and lower side well developed; dorsal and anal fins connected to the tail by a distinct membrane. Dark brown with irregular dark blotches and a black spot on the upper pectoral fin. Atlantic, English Channel, North Sea, western Baltic and Mediterranean. Common. (**XI, 31**)

Mola mola (Linnaeus): Sunfish. Length up to 3 m. Almost circular, heavy body with dorsal and anal fins similar in shape and size and situated at the end of the body; tail short. Small mouth with beak-like tooth in each jaw; small eye. Back grey or grey-brown; sides and belly lighter; fins dark. Open ocean species usually seen floating near the surface, often on its side, in warm weather. Atlantic, western English Channel and Mediterranean. Uncommon. (**XI, 30**)

Class: Reptilia (Reptiles)—Turtles

Dermochelys coriacea (Linnaeus): Leathery turtle. Length up to 1·8 m over the curve of the shell; weight, 350–500 kg when adult. 5 or 7 prominent, longitudinal ridges on carapace, these sometimes notched; carapace covered by thick leathery skin.

Upper jaw has 2 distinct tooth-like points at front. Black or dark brown, usually with lighter flecks. Usually seen well offshore feeding on salps and jellyfish. Atlantic and Mediterranean. Essentially warm-water species occasionally appearing in north-west Europe; known to breed in Mediterranean. (p. 148)

Caretta caretta (Linnaeus): Loggerhead turtle (p. 148). Length up to 1·1 m over the carapace, usually smaller. Shell with 5 plates down each side; horny, oval and elongated; young may have 'saw-backed' appearance due to projections on central row of vertebral plates. Red-brown in colour, young have dark streaks. Young may be confused with the rare Kemp's ridley turtle (*Lepidochelys kempii*) but this species has a less elongated carapace. In deep water but may be close to shore where it feeds on crabs, sea-urchins and molluscs as well as jellyfish, etc. Atlantic and Mediterranean. Usually only the young found in north-west European waters—these probably drift from the Gulf Stream. Breeds in Mediterranean.

Chelonia mydas (Linnaeus): Green turtle. Length of shell up to 1·4 m. Oval shell similiar to loggerhead but with only 4 plates down each side. Shell usually brown or olive with darker mottling, streaks and blotches. The scales on top of the head have light borders. Usually found in warm shallow waters where there is plenty of seaweed. Extremely rare in European waters but occasionally turns up near Atlantic coasts and in the Mediterranean.

Eretmochelys imbricata (Linnaeus): Hawksbill turtle. Length of shell up to 90 cm. Easily distinguished from other turtles by overlapping horny plates on the carapace. Adults have narrower shells than the young animals. A tropical species, extremely rare in European waters but may appear occasionally.

Class: Aves (Birds)

Gavia immer (Brunnich): Great northern diver. Length 75 cm. In summer glossy black head and neck with white bands; black back with white spots. In winter dark above and white below. Heavy black bill. Sexes alike. Call: a loud wail and yodelling quack. In coastal waters diving for fish and benthic organisms. Nests in Iceland, Greenland and North America on lakesides. Atlantic, English Channel and North Sea. Rare. (p. 148)

Gavia stellata (Pontoppidian): Red-throated diver. Length 60 cm. Red throat in summer with rest of neck grey; back and wings brown; under parts pale; sexes alike; red throat patch lost in autumn. Call: a wail, 'kwuk-kwuk-kwuk' in flight. Found in coastal waters and also inland on fresh water. Breeds in northern Europe, Scotland and north-western Ireland; eggs laid on the ground or on a heap of weed. Common in breeding range. (**XII, 15**)

Fulmarus glacialis (Linnaeus): Fulmar. Length 45 cm. Dark grey back and wings; white head, tail and under parts; tubular nostrils; short tail; stiff flight. Sexes alike. Oceanic bird nesting on cliffs. Atlantic south to Finisterre, English Channel and North Sea. Common. (**XII, 7**)

Puffinus puffinus (Brunich): Manx shearwater. Length 35 cm. Black head, neck and upper parts; white chin, throat and under parts. Sexes alike. Call: a coo. Low rapid flight. Oceanic bird sometimes gathering in 'rafts' on the sea. Nests in a burrow excavated by parents or in a disused rabbit-hole. Atlantic, English Channel, North Sea and Mediterranean. Common. (**XII, 18**)

Hydrobates pelagicus (Linnaeus): Storm petrel. Length 15 cm. Black plumage; white rump, long wings; square tail; slim bill. Sexes alike. Call: a purring sound (at night on breeding ground). Low wavering flight, often paddling the water with its feet. Oceanic bird nesting in burrows dug by parents or in disused rabbit-holes. Atlantic, western English Channel, northern North Sea and Mediterranean.

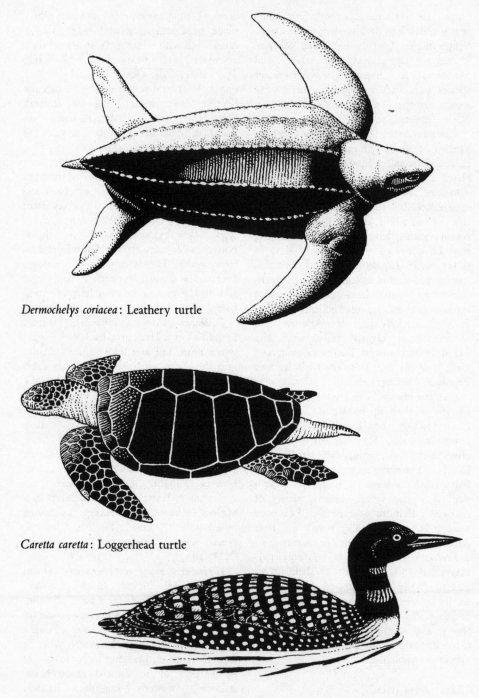

Dermochelys coriacea: Leathery turtle

Caretta caretta: Loggerhead turtle

Gavia immer: Great northern diver

Common; may occur inland after storms. (**XII, 19**)

Diomedea melanophris Temminck: Black-browed albatross. Length 79 cm. Long wings with a spread of over 2 m; unmistakable because of its large size. Glides with stiff wings but can fly powerfully. Very rare visitor to European waters from the southern hemisphere. The yellow-nosed and grey-headed albatrosses are even rarer visitors and do not possess the yellow bill and dark eye streaks of the black-browed albatross.

Sula bassana (Linnaeus): Gannet. Length 90 cm. White plumage; black wing tips; pale blue bill; buff-coloured head and neck. Characteristic body tapered at both ends in flight. Sexes alike. Young dark brown with white cheeks until 4 years old. Call: a screech. Offshore bird seen gliding low over the sea and diving for fish. Nests in large colonies on cliff ledge or flat ground. Atlantic, English Channel, North Sea and western Mediterranean. Locally common. (**XII, 17**)

Aythya marila (Linnaeus): Scaup. Length 45 cm. Drake: dark green head, black breast and light grey back. Duck: dark brown with large white face patch. Drake duller in colour July–November. Usually silent; duck may give harsh 'karr-karr-karr'. In bays and estuaries diving to bottom to feed, nests in northern Europe, Iceland and occasionally northern Scottish Islands. Atlantic, English Channel, North Sea, Baltic and northern Mediterranean. Common locally. (**XII, 20**)

Somateria mollissima (Linnaeus): Eider duck. Length 60 cm. Drake: white above, black below but in eclipse plumage (July–October) resembles a dark female. Duck: brown with white on wing. No forehead. Call: 'ah-oo'. Usually near rocky shores. Nests on shore. Atlantic north of Ireland, northern North Sea and Baltic. Common. (**XII, 23**)

Anas penelope Linnaeus: Wigeon. Length 45 cm. Drake's head chestnut coloured with a buff forehead and crown; upper parts grey; green wing patch. July–October resembles the duck. Duck has a green wing patch and is mottled brown above to buff on the lower parts. Call: a high descending whistle. Nests on the ground among heather or bracken in northern Europe and Scotland. Atlantic, English Channel, North Sea, Baltic and Mediterranean. Common. (**XII, 21**)

Clangula hyemalis: Long-tailed duck

Melanitta fusca: Velvet scoter

(The long string of all-black ducks, flying low, are the common scoter.)

Clangula hyemalis (Linnaeus): Long-tailed duck. Length 50 cm. Drake: long pointed tail; dark brown upper parts and breast, white flanks and belly; head and neck white in winter and brown with white on side of the face in summer. Duck: short tail, sides of head white, brown upper parts and white beneath. Short bill. Call: a yodelling whistle. Along coasts and for short distances out to sea. Nests on the edge of northern lakes. Atlantic north of the English Channel, North Sea and Baltic. Common. (p. 149)

Melanitta nigra (Linnaeus): Common scoter. Length 50 cm. Drake: black with orange mark on bill. Duck: dark brown with buff on cheeks. Call: a high piping. In coastal waters diving for shellfish etc., often in flocks of a dozen or more. Nests on the shore. Atlantic, English Channel, North Sea and Baltic. Common. (**XII, 22**)

Melanitta fusca (Linnaeus): Velvet scoter. Length 56 cm. Drake is black with white spot near the eye and white wing-bar. Duck is dark brown with 2 white face patches. Nests in hollow in the ground in the Baltic area. A sea duck occurring in small groups, often diving in synchrony to feed on mussels, shrimps and crabs. English Channel, North Sea and Baltic. Basically a northern bird, not so far-ranging as the common scoter. (p. 149)

Mergus serrator Linnaeus: Red-breasted merganser. Length 56 cm. Drake: dark green head with double crest; chestnut breast; grey and black back. Duck: brown and grey with white wing-bar. Call: a short, quiet 'quack'. Along coasts and estuaries particularly during winter. Nests on the shore under bushes or among rocks. Atlantic, English Channel, North Sea, Baltic and northern Mediterranean. Common. (**XII, 24**)

Pandion haliaetus (Linnaeus): Osprey. Length 58 cm. Dark brown upper parts; white under parts speckled brown; dark brown band on side of white head; long wings with distinctive angle in flight. Sexes

alike. Call: a series of shrill, cheeping whistles. In estuaries and sheltered lochs as well as fresh water; flying slowly, soaring, hovering and diving, feet first, for fish. Nests in trees. Scotland, Baltic and southern Mediterranean. Rare in Britain, less so on mainland Europe. (**XII, 29**)

Falco tinnunculus Linnaeus: Kestrel. Length 34 cm. Male has black spotted chestnut upper parts, bluish-grey head, rump and tail; tail has a black terminal band; under parts buff with dark streaks and markings. Female is rufous brown with blackish barrings above; pale below with dark streaks; tail barred. Pointed wings. Characteristically seen hovering inland and near rocky coasts. Nests on cliffs, in trees or ruins. Atlantic, English Channel, North Sea, Baltic and Mediterranean. Common. (**XII, 26**)

Stercorarius skua (Brunnich): Great skua. Length 50 cm. Brown plumage; white wing patch; short tail. Has the appearance of a heavily built gull. Sexes alike. Open sea; often seen soaring with gulls. Nests in bog or heath on Orkney, Shetland and Iceland. Atlantic, northern North Sea, western Mediterranean. Not uncommon but very local. (**XII, 27**)

Stercorarius parasiticus (Linnaeus): Arctic skua. Length 45 cm. Dark brown upper parts, under parts lighter; in flight 2 distinctive, central tail feathers project in a point. A dark and a light form exist. Sexes alike. Call: a wailing 'ka-aaow' or 'tuk-tuk'. Nests colonially on moors and tundra in northern Scandinavia, central Baltic, Iceland and northern Scotland. Could be near any coast during migration periods. Fairly common. (**XII, 25**)

Larus canus Linnaeus: Common gull. Length 40 cm. Grey back, white under parts; triangular black wing tip with a white spot; yellow green bill and legs. In winter the head is grey. Sexes alike. Young is dark brown above; pale below. Call: a high-pitched 'kak-kak-kak' and a scream. On coasts and inland. Nests in small colonies on the ground. Atlantic, English Channel, Baltic and Mediterranean. Less common than the name implies but not rare. (**XII, 6**)

Larus argentatus Pontoppidian: Herring gull. Length 58 cm. Grey back; white under parts; yellow bill with red spot; pink legs. Sexes alike. Immatures brown. Loud, clear call. On coasts and inland. Nests in colonies on ground, cliffs or buildings. Atlantic, English Channel, North Sea, Baltic and Mediterranean. Very common. (**XII, 4**)

Larus fuscus Brehm: Lesser black-backed gull. Length 53 cm. Dark grey back; white under parts; yellow legs. Sexes alike. Immatures brown with pale under sides. Call: loud and varied. Often seen inland as well as on coasts. Nests in colonies on ground or on cliffs. Atlantic, English Channel, North Sea, Baltic and Mediterranean. Common. (**XII, 1**)

· *Larus marinus* Linnaeus: Greater black-backed gull. Length 68 cm. Black back; white under parts; pinkish legs; large heavy bill. Sexes alike. Young are brown above and pale on undersides. Call: a low pitched 'kow-kow-kow'. Coastal, rarely seen inland. Usually nests in small groups on ground or on rocky cliffs. Atlantic, English Channel, North Sea, Baltic and Mediterranean. Common. (**XII, 3**)

Rissa tridactyla (Linnaeus): Kittiwake. Length 40 cm. White with grey wings and back, and triangular black wing tips; short black legs; yellow-green bill. Sexes alike. Young have a distinctive dark stripe along the wing. Call: characteristic 'kitt-ee-wake'. Oceanic bird nesting in colonies on rocky cliffs. At sea flies low over the water often in lines. Atlantic, English Channel, North Sea and Mediterranean. Common. (**XII, 2**)

Sterna hirundo Linnaeus: Common tern. Length 35 cm. White with black head and nape; red bill usually with black tip. Sexes alike. Call: a harsh 'kee-urr'. Often seen diving for fish close to the shore: sometimes

in large groups. Nests on shingle beaches, salt marshes or small islands. Atlantic, English Channel, North Sea, Baltic and Mediterranean. Very common. (**XII, 9**.) The Arctic tern (*Sterna paradisaea* Pontoppidian) is very similar but slightly larger with an all-red bill and shorter legs. Breeding areas more northerly than common tern.

Sterna dougallii Montagu: Roseate tern. Length 38 cm. White with black crown and nape; pink tinge on breast in summer; black bill which is red at the base in summer; red legs; long white tail streamers. Sexes alike. Call: soft 'chivy' or coarse 'z-a-a-p'. Often seen fishing close to shore; never inland. Nests in colonies in sand or shingle. Atlantic, English Channel and North Sea. Local. (**XII, 8**)

Sterna albifrons Pallas: Little tern. Length 23 cm. White forehead with black crown in summer, grey in winter; yellow bill with black tip; orange feet. Sexes alike. Call: rapid cries of 'kik-kik' or 'pee-e-eer', rather noisy. Inshore. Nests in small colonies on the beach. Atlantic, English Channel, North Sea and Mediterranean. Decreasing and local. (**XII, 5**)

Sterna sandvicensis Latham: Sandwich tern. Length 40 cm. Black crown and crest streaked white in winter; black bill with yellow tip; grey wings and back. Sexes alike. Young brown and white. Call: loud and harsh, 'karrick', 'kirwhit'. Fishes offshore often out of sight of land. Nests in colonies in sandy or shingly areas. Atlantic, western English Channel, North Sea and western Baltic. Common. (**XII, 11**)

Alca torda Linnaeus: Razorbill. Length 40 cm. Heavy bill; black upper parts; white wing stripe. In summer head and neck black with white line from bill to eye; in winter throat and sides of neck white. Sexes alike. Oceanic bird seen flying low over waves in small flocks or floating and diving for up to 50 seconds at a time. Lays eggs on rocky cliffs. Atlantic, English Channel, North Sea, western Baltic and western Mediterranean. Locally common. (**XII**, 10)

Uria aalge (Pontoppidian): Guillemot. Length 41 cm. Pointed bill; upper parts dark brown in summer; more grey in winter. Occasional 'bridled' specimens have white ring around the eye and a white line running back over the sides of the head. Sexes alike. Call: 'arrr'. Oceanic species coming ashore to lay eggs on rocky cliffs. Atlantic, English Channel, North Sea, western Baltic and western Mediterranean. Common. (**XII**, 12)

Cepphus grylle (Linnaeus): Black guillemot. Length 32 cm. In summer plumage black with conspicuous white wing patch; in winter upper parts barred black and white and head and under parts white. Red legs. Sexes alike. Twittering call. Flies low over the water. Less oceanic than the guillemot, remaining closer to land throughout year. Lays eggs on rock cliff or among boulders. Atlantic, south to southern Ireland, northern North Sea and Baltic. Common. (**XII**, 13)

Fratercula arctica (Linnaeus): Puffin. Length 30 cm. Characteristic large, blue, yellow and red bill in summer and horn-coloured in winter. Upper parts black, under parts white. Sexes alike. Unmusical guttural call. Flies fast and low in short lines. Oceanic bird moving far out to sea except when coming ashore to nest in disused rabbit-hole or burrow dug by parents. Atlantic, North Sea and western Mediterranean. Common. (**XII**, 14)

Pyrrhocorax pyrrhocorax (Linnaeus): Chough. Length 38 cm. Black plumage with greenish-blue gloss; curved red bill; red legs. Sexes alike. Call: a high-pitched 'kiah' or 'k'chuff'. Usually seen close to steep rocky cliffs and inland among high mountains. Nests in holes in cliff face or in sea caves. Atlantic coasts. Rare and local. (**XII**, 28)

Sturnus vulgaris Linnaeus: Starling. Length 22 cm. Iridescent green, blue and purple plumage; white or buff spangling in winter. In flight characteristic short tail and pointed wings. Typically a town bird but northern and eastern European birds migrate to Britain in winter and large flocks may be seen at sea in the English Channel, southern North Sea and Baltic. Very common. (**XII**, 30)

Class: Mammalia (Mammals)
Order: Cetacea (Whales)

Balaenoptera acutorostrata Lacépède: Lesser rorqual or pike whale. Up to 10 m long. Stout body; broad, low snout; back fin in hind $1/_3$ of body with concave trailing edge; tail flukes with notch in middle of hind edge; numerous parallel grooves on throat and chest. Black above, pure white below; conspicuous white patch on outer surface of flipper. Oceanic species. Atlantic, western English Channel and northern North Sea. Relatively common. (**XI**, 34)

Globicephala melaena (Traill): Pilot whale or blackfish. Length up to 9 m. Long slender body; bulging 'forehead' and very short beak. Long back fin pointing backwards in middle of body; long, narrow, pointed flippers; notch in middle of hind edge of flukes. Most of body black with white patch behind the chin which may stretch in a streak along the belly. Usually in schools; may come close to the shore. Atlantic, English Channel and North Sea. Common in northerly areas in the vicinity of Orkney and Shetland and not infrequent further south. (**XI**, 33)

Orcinus orca (Linnaeus): Killer whale or grampus. Male may attain a length of 10 m, possesses a characteristic angular dorsal fin up to 2 m high. Females, up to 7 m in length, do not possess the characteristic dorsal fin. No beak; 10–13 massive teeth on each side of upper and lower jaws. Both sexes conspicuously marked in black and white. A ferocious species feeding on squid, fish, seals and other whales. May be dangerous to man. Atlantic. Not uncommon. (**XI**, 37)

Phocaena phocaena (Linnaeus): Common or harbour porpoise. Length up to 2 m. Evenly rounded snout with no beak; low, triangular back fin in middle of back; oval flippers with rounded tip; tail flukes have a notch in the middle of the posterior margin. Black back, white belly, grey sides. Occurs singly or in pairs, never in schools. Often seen inshore and even ascends estuaries occasionally. Atlantic, English Channel and North Sea. Common. (**XI, 36**)

Tursiops truncatus (Montagu): Bottle-nosed dolphin. Length up to 4 m. Robust body; well-defined snout; moderately high dorsal fin in the middle of the back pointed backwards. Back and sides black or dark brown, throat and belly white. Occurs in schools and will often ride the bow wave of larger boats. Atlantic, English Channel, North Sea, Baltic and Mediterranean. Common.

Delphinus delphis Linnaeus: Common dolphin. Length 2–2·5 m. Slender and graceful with a long, narrow, well defined snout. Back fin on middle of body with tip directed backwards and concave posterior edge. 40–50 teeth on each side of upper and lower jaws. Dark colour on top of back; black flippers from which a dark streak extends forward to the lower jaws; a black circle surrounds the eye and a black streak extends from this to the base of the beak; streaks of light and dark pigment along the sides. Atlantic, English Channel, North Sea (except central region) and Mediterranean. Common. (**XI, 35**)

Part Two

CHAPTER FOUR

Seamanship and Longshoremanship

The sailor/naturalist has to be both a good seaman and a good longshoreman. The side of seamanship that will be tested is the ability to land on inhospitable shores and to anchor securely in confined spaces. A longshoreman is a waterman who makes his living by the shore. He is out in all weathers, is a master of the small boat, a local pilot and, therefore, able to find his way home in bad conditions. The yachtsman/naturalist should aim to combine the skills of the two. This chapter confines itself to these peculiar needs, leaving pure seamanship and navigation to books devoted to those subjects.

LANDING IN REMOTE AND EXPOSED PLACES

It is not always possible to anchor when wishing to land on an island or remote headland. The chart should be inspected with care, having noted the state of the tide, the rise and fall, the direction of the wind and any expected changes gleaned from the weather forecast or observation.

The chart will show the dangers to be avoided and it is worth ringing the serious ones with a pencil so they are immediately recognisable. The depths and composition of the bottom should also be noted. From this information it will be possible to select a landing place and perhaps a temporary anchorage. 'Landing place' is often marked and the chart should be closely examined for such information. Small stone piers may be badly delineated yet show up when viewed with a keen eye or chart magnifying glass.

However, the chart does not hold the monopoly of information and careful observation with binoculars may confirm suspicions or reveal problems. Swell, for example, may come round from an island's weatherside. This may cause a surge at the landing place or even breakers on the beach. If an island is small, or very exposed, the swell may arrive at differing times from both headlands, creating a confused and uncomfortable sea under the island's lee.

LYING OFF

Some yachts, particularly those of heavy displacement with a long straight keel, will heave-to comfortably. The headsail is backed, counteracting the driving effect of the main or mizzen, and the helm is lashed down. The yacht will still be subject to tide and will make leeway, so it is important to see that there are no dangers close down tide or to leeward. If the yacht has an auxiliary and the stay is to be a short one, the engine should be running.

The crew must be competent to do two main jobs—to handle the yacht short-handed and to man the dinghy skilfully—otherwise attempting to land in exposed and remote areas is foolhardy.

It is important to establish a plan that anticipates problems beforehand, rather than responding afterwards to an emergency. As there is unlikely to be more than one dinghy, the dinghy must be carefully looked after and it may be prudent to leave one of the landing party on the beach while the others explore. The dinghy man not only looks after the boat, but acts as the link between the shore and the yacht. The dinghy man should display a visual signal, perhaps a coloured flag or burgee on a stick, which he plants in a place conspicuous to the yacht, indicating that all is well ashore. This signal reassures those afloat. If the flag is removed, those on the yacht should watch the shore carefully.

Equally, the yacht may require the shore party to return immediately. They may have heard a weather forecast, or realised that conditions were becoming unfavourable. A visual signal, either selected from the international code (P meaning 'In harbour all persons should report on board as the vessel is about to proceed to sea'), or an agreed sign, such as a white flag, perhaps a tea towel, flying from the cross trees on the side of the mast not blanketed by the sail, indicating the same thing. Sound and light signals are also useful to call attention to the visual signal, and are of assistance if visibility deteriorates. The dinghy man also should be able to call his companions' attention to messages from the yacht.

Some yachts carry small portable radio telephones and these are very useful. The inexpensive citizens' band models that used to be available are frowned on by most European authorities, who have yet to see the undoubted convenience and safety value of such a frequency and equipment. However, it is not sensible to rely completely on even the most sophisticated and licensed transmitter/receiver and the arrangements described above will be useful to the well-equipped, the radio acting as an additional aid.

Even in the remotest spots, flares and pyrotechnics should only be used when outside help is urgently required, or much voluntary and official time and effort may be spent to no purpose, following reports of distress signals.

ANCHORAGES AND ANCHORS

Considerations and equipment for the temporary anchorage are discussed first. Lying to, or gilling about in a motor cruiser, under the lee of the land, requires constant activity and attention on board. This will reduce the amount of time available for observing the activities of those ashore. The kedge anchor can remove a great deal of that anxiety and its employment is often much neglected.

The word kedge comes from 'catch', meaning to catch the ground, though some American writers use the word for a 'fisherman anchor'.

Before the auxiliary, or indeed the steam, engine, shipping used to kedge to await the tide. The Downs in the Channel Narrows used to be full of ships awaiting a change in tide or weather. It was common to anchor in deep water without a vestige of shelter, for the purpose was to retain an advantage, rather than a quiet place for the night. The practice was known as 'tiding over'. Perhaps the supreme masters of the kedge are Danish seiners, fishing boats not exclusively from Denmark, who use the anchor for securing one end of the seine, steaming in a circle to net the fish and recover the anchored seine warp. The envelope of net is then pulled aboard.

The kedge is defined as a small anchor commonly used with a warp rather than a chain. The principal or bower anchor, which even today usually employs chain, is the vessel's main anchor and she may have at least two bower anchors to one kedge. The principal requirement of a kedge is that the anchor is convenient to use, as well as being effective for the size of the yacht that carries it.

Experienced sailors may argue that in exposed conditions the main anchor should be used in preference to the kedge and they have reason on their side. The decision depends on the ability to handle the main anchor and cable, most probably chain, at short notice and quickly. With yachts equipped with a powerful windlass of proven reliability, this is the right decision. Otherwise, the lighter kedge has the advantage of quick handling and if the anchor becomes fouled and has to be buoyed and left, there is not the same element of loss as with

Fig. 9.   Anchors and their parts.

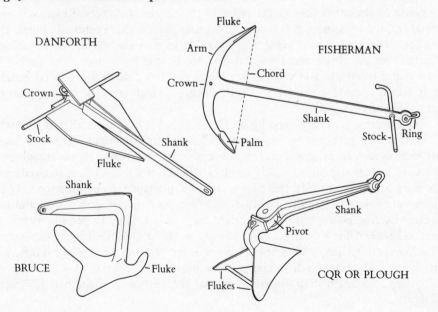

DANFORTH

FISHERMAN

Fluke

Arm

Crown

Chord

Crown

Shank

Crown

Stock

Shank

Stock

Ring

Stock

Fluke

Shank

Palm

Shank

Fluke

Pivot

Shank

BRUCE

Fluke

Flukes

CQR OR PLOUGH

a main anchor. Kedge anchors must, therefore, be more convenient to weigh and bring aboard. It is important, however, that the right anchor should be used for the right sub-stratum, as some anchors perform better than others on different types of bottom. There is an endless choice of anchors as this list of some of the more common names suggests: Baldt, Benson, Bruce, Colin, C.Q.R., Danforth, Digger, Fisherman, Gillois, Grapnel, Herreshoff, Martin, Marrel, Narspel, Navy, Nicholson, Northill, Mushroom, S.A.K., Stokes, Viking and Wishbone, to which can be added all the ancient wooden and stone hooks from Greece to Chiloe. Most yachts would founder if they carried anything like this selection and so there are fads and fancies. A short list notable for good holding in exposed situations would include four of the above and a strong bias for one type for the kedge. It is worth looking at these four in some detail. In alphabetical order they are the Bruce, C.Q.R., Danforth and Fisherman.

The Bruce Anchor is a recent invention and was the result of the great anchor development period following the North Sea oil bonanza. Huge Bruce anchors secure many of the semi-submersible rigs and form the moorings in the later building stages of the concrete oil production platforms. The anchor is said to work well with less cable or warp than other alternatives. There are no working parts; it has absolute roll stability and so the yacht can veer through 360 degrees without the anchor breaking out. The three points of the fluke are said by the makers to give 'high rock hooking capacity'. There is no doubt that the Bruce is a formidable advance in anchor thought.

The C.Q.R. or Plough Anchor was invented by Professor Sir Geoffrey Taylor, F.R.S. The initials are shorthand for 'secure'. The anchor makes use of two ploughshares that pivot to a degree on the shank. The anchor is designed to roll to one side or the other and as the point of the shares is slightly downward and weighted, the horizontal pull starts the digging process that eventually buries the anchor. It is almost impossible for the cable to foul the shank, an advantage shared with the Bruce and Danforth but not by the Fisherman. The anchor is now out of copyright and anchors have been developed using this general design. It is worth obtaining the true C.Q.R. anchor, which is made by Simpson & Lawrence of Glasgow.

The Danforth was developed by R. D. Ogg and R. S. Danforth and with the C.Q.R. is most favoured by yachtsmen in Europe and America. The broad flukes are close to the central shank and the stock is across the crown. Its advantage over the C.Q.R. is that it stows neatly on deck, though if the C.Q.R. is secured over the stem or bowsprit roller, this edge disappears, perhaps even making the C.Q.R. more convenient to handle. The hi-tensile version employs special steel and has greater holding power than the standard, as it is less likely to bend or break. It digs in quicker due to sharpened flukes but is also a good deal more expensive. The Danforth Utility is sometimes known as the Northill anchor and is different in design. The Danforth anchor, in common with the C.Q.R., is out of copyright, but again it is worth seeing that the anchor is of the true Danforth design.

The Fisherman Anchor is the traditional anchor and is used effectively all over the world. Claud Worth, the pioneer British cruising yachtsman, specified that the flukes should be sharp and broad, so that they dig in easily and once there hold. The flukes should subtend an angle of 40 degrees with the shank, which should measure from crown to shank hole at least $1^1/_2$ and not more than $1^2/_3$ times the length of the chord, or the distance between the fluke points. The stock, of course, is longer than the chord and this makes the anchor roll over, so that a fluke is encouraged, by horizontal pull and weight, to pierce the ground. A light Fisherman is not worth stowing, for the prime reason for carrying this type of anchor at all is its proven ability to pierce the ground when other more convenient and lighter anchors fail. It is better than the Danforth and has a slight edge over the C.Q.R. in kelp- and weed-covered rocky areas. The Fisherman anchor does not bury itself in the way of the C.Q.R. and is, therefore, referred to as a surface anchor as opposed to a burying type. The drawback of the Fisherman is the difficulty of bringing it aboard conveniently in a seaway without a cat head or a davit. A way of overcoming this is to attach a line to the crown before anchoring and secure this, on retrieval, to the jib halyard, one crew member hoisting away while the other ensures that the anchor does not foul the ship, lifelines or rigging.

A glance at the table below is useful in measuring performance but the decision of which anchor or anchors to use comes from past experience and bias.

The Bruce, C.Q.R., Danforth and Fisherman all have a place on board. Note how the C.Q.R. anchor is lighter than the standard Danforth though heavier than the Danforth hi-tensile. Eric Hiscock remarks in his classic book *Cruising*

Anchor Weights, Warp and Chain Sizes for Storm Conditions

Length of yacht in metres	Bruce		Danforth hi-tensile		Danforth standard		C.Q.R.-Plough		Fisherman		Minimum warp & chain lengths & size	Minimum chain cable lengths & size
	kg	lb	kg	lb	kg	lb	kg	lb	kg	lb	100m × $^1/_2$" dia. + 4m $^5/_{16}$" ch.	65m × $^1/_4$" ch.
6–10	10·0	22	5·4	12	5·9	13	6·8	15	18·2	40		
10–11	10·0	22	5·4	12	10·0	22	9·0	20	20·4	45	100m × $^5/_8$" dia. + 6m $^3/_8$" ch.	65m × $^5/_{16}$" ch.
11–13	20·0	44	9·0	20	18·2	40	11·3	25	22·7	50	150m × $^5/_8$" dia. + 8m $^3/_8$" ch.	75m × $^3/_8$" ch.
13–16	20·0	44	15·9	35	29·5	65	15·9	35	31·2	70	150m × $^5/_8$" dia. + 8m $^3/_8$" ch.	100m × $^7/_{16}$" ch.
16–20	30·0	66	27·2	60	38·6	85	20·4	45	54·5	120	200m × $^3/_4$" dia. + 8m $^3/_8$" ch.	100m × $^1/_2$" ch.

Note: a kedge anchor is normally two-thirds the weight of the bower.

Under Sail that he does not consider it wise to use any patent anchor under 30 pounds, for although it may have the necessary holding power, the anchor may not be heavy enough to force its way through weed and if it does not bite at one, it will foul with weed and not bite again.

The Warp

The amount of warp or chain for anchoring depends on the weather and the maximum depth of water expected. It must include the rise and fall of the tide, as well as the distance the yacht's fairlead is off the water. The length of cable divided by the figure for the maximum depth is known as the scope and is all important. When anchoring with chain alone, at least three times the depth is called for, while with a nylon and chain warp, as specified in the above table, eight to ten times the depth would suit storm conditions.

The nylon warp may be either laid or braided. Laid nylon rope stretches and is 15–25 per cent more elastic than manila. Braided line is about 10 per cent less elastic than laid, but is still suitable for anchoring. It has the great advantage of not kinking and so can be stowed without a struggle.

A small length of chain should be connected to the nylon warp with an eye splice armed with a thimble. The short lengths of chain suggested in the table not only ensure a horizontal pull on the anchor, but cut down chafe, as the warp slides about the bottom, in response to the pull of the yacht.

There are at least three ways of further increasing the holding power of the anchor. The first is the use of a weight shackled to the join between chain and warp. Once the yacht responds to the wind or strong tide, the weight is lifted. This prevents or reduces snubbing, which may break out the anchor or cause damage to the sampson post or windlass. A traveller, such as a large 'D' shackle, can be used on the warp or chain to lower the weight after the anchor is down, but this may chafe the rope if used for an extended period without attention.

Another way is to adopt a mooring technique. A large plastic buoy is attached to the anchor warp or chain and this floats, taking care, particularly in the case of chain, of the majority of the weight. A short length, say half the overall length of the boat, connects the buoy to the yacht. As the wind and sea rises, the vessel pulls and in so doing, exerts pressure on the buoy, trying to sink it. If the buoy is of proper size, this will be a difficult job and take much of the energy out of the weather. By supporting the chain, the buoy relieves the bow of the yacht, allowing easy response to each wave.

The third gilhick may be used when anchored with chain cable alone. A short length of stout nylon is introduced between the sampson post and the chain. This will absorb much of the shock and add to both safety and peace on board. The nylon is permanently shackled to the yacht for ready use. It then may be tied to the cable using a chain hitch, or by a Cornish hook that goes through, or better still across, the link, in the manner of a cable stopper or devil's claw.

Buoying the Anchor

There are ways of recovering anchors, but prevention is better than cure and the best precaution is to buoy. A non-floating line is secured to the crown of the anchor and buoyed. If the anchor refuses to respond to normal weighing, hauling on the crown will usually ensure recovery.

A buoyed anchor can be a nuisance, though, in crowded anchorages, as the line may foul propellers. A line that floats is a real menace. Anchor buoys should, therefore, only be used when the ground is thought to be foul. Then it is useful to use more line than normally necessary and retrieve the buoy aboard.

There are other ways of recovering a recalcitrant anchor without the trip line. When the flukes have caught a mooring chain, for example, a grapnel with a retrieving line may be used to lift the chain. The yacht's anchor is then released, allowing it to drop free. If the grappled chain is not too heavy, the links may be brought to the surface, secured with a line from the bow and both the grapnel and the anchor dropped away and freed with ease.

Other fouled anchors may be cleared by sailing or motoring them out. Under wind alone, the sails are set and the sheets hauled in tight, the foresail is backed to give the yacht a flying start. As the yacht tacks over the anchor, the chain is hauled aboard and secured when at its shortest. The sudden snatch should break out the anchor. Working the anchor out under power works in the same way, except that the need to tack to an 'up and down' position is dispensed with. The cable is hauled in by windlass or hand, perhaps with the aid of the engine, and snatched out in the same way. It is important to secure the chain or warp to a strong point first, rather than leaving on the locked gypsy or windlass drum, as this may cause damage. Both methods require care, for many a cruising yachtsman has a shortened finger as a result of anchor work.

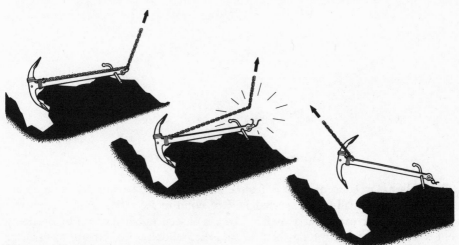

Fig. 10. Scowing an anchor.

In particularly vulnerable areas, it is worth considering a method known in America as 'scowing'. The cable is unshackled from the ring and made fast to the crown, leading the chain back along the shank. The cable is stoppered or fastened to the ring with spun yarn. When the cable is shortened, the angle increases, eventually putting pressure on this lashing, which gives way and the anchor breaks out readily (see Fig. 10).

The Danforth Sure-Ring anchor achieves the same idea with a slotted shank so that the ring can slide towards the crown, so breaking out the anchor when the yacht exerts a full forward pull.

Using Two Anchors

Two anchors in more peaceful but potentially challenging surroundings, give great security and, perhaps as important, greater peace of mind. The lines should subtend an angle of 45 degrees (see Fig. 11). This stops the yacht charging about and enables the vessel to be kept clear of dangers that might well be within the circle of one anchor. It is important to ensure that there is equal strain on each anchor by adjusting the scope individually.

Fig. 11.   Riding to two anchors.

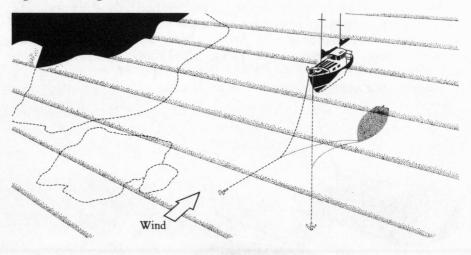

Wind

Mooring in Inlets and Rivers

The yacht has to be secured away from either bank, whether the tide is going out or coming in. Equally, an expected shift of wind in a confined inlet makes this form of mooring particularly useful. The vessel secures her position by dropping one anchor, while stemming the tide, and then allowing the stream to take the yacht astern twice the scope of the proper length of cable. The second anchor is

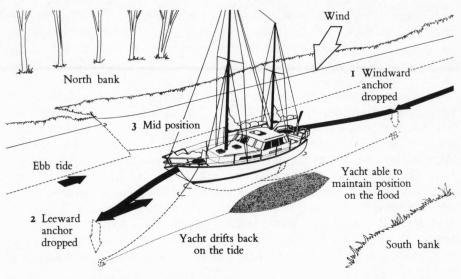

North bank

Wind

1 Windward
anchor
dropped

3 Mid position

Ebb tide

Yacht able to
maintain position
on the flood

2 Leeward
anchor
dropped

Yacht drifts back
on the tide

South bank

Fig. 12. Anchoring in narrow waters.

then let go and the yacht hauled back to the mid point between both anchors (see Fig. 12).

In Scandinavia, where the rise and fall of the tide is of little moment, yachts make best use of those corrugated and island-speckled coasts. In western Sweden, sailors have devised a method of pinning the bow to ice-smoothed rocks with a

Fig. 13. Pinned to a rock.

Steel spike

Wind

Kedge anchor

steel spike. In calm weather, with enough breeze to keep the yacht clear, the stern is allowed to float free, the mizzen positioning the yacht so as to take best advantage of the lee of the rock. Care should be taken to see that there are no hidden dangers, and a stern anchor may be employed to ensure the yacht does not swing onto them (see Fig. 13).

In western Norway, the yachtsman has developed techniques that can be useful in areas with a greater rise and fall. The yacht is anchored by the stern, head to wind, anticipating any change of direction by the proper positioning of the stern anchor, with the bow secured to a rock or tree. If a suitably steep-sided place is found, selected by the lie of the land and a glance at the weed fringing of the rocks, it is possible to leap from the bow or bowsprit without resort to the dinghy. The yacht is secured a safe distance from the shore by hauling on the stern anchor warp. In reasonably sheltered places, further lines from the bow and another anchor out astern will ensure that the yacht is held in the right position in almost any eventuality. There are few more pleasant places to be than in a rocky inlet tied to a tree. The Scandinavians have developed this to a particularly fine art (see Fig. 14).

Fig. 14. Mooring Scandinavian fashion.

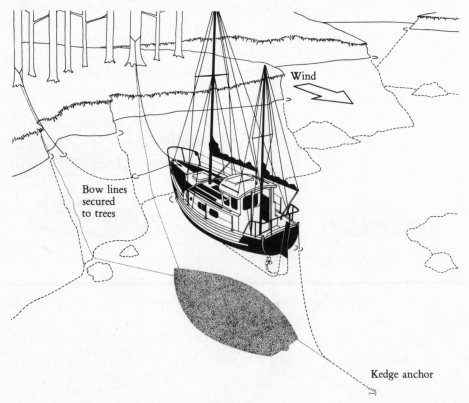

Wind

Bow lines
secured
to trees

Kedge anchor

THE ART OF DRYING OUT

The art of drying out is poorly covered in sailing literature. So much has been written and said about keeping the yacht afloat that to turn the mind to grounding on purpose is too paradoxical for the normal salt-water author. However, these techniques have application to the yachtsman/naturalist, though the designers of racing yachts who see them also as suitable for cruising have not helped much for the fin and skeg type of design is not always suitable for supporting a hull against a wall and can be dangerous if used with legs. Also, a sharp keel sinks easily into the ground and can cant the yacht at an awkward angle, exerting destructive strains, and the skeg rudder is usually unprotected and particularly vulnerable. These yachts are built for use in water and the yachtsman/naturalist has greater demands than that.

On the other hand, the last decade or so has seen a great increase in twin-keeled yachts which are ideal for exploring corners of the sea environment. They need ordinary care in taking the bottom. The chart usually provides general information about the type of bottom and a glance at the shore will help reinforce this. The echo-sounder, too, will show whether the ground is smooth and gently sloping or corrugated with weed and rock echoes. A paper recorder will make this more obvious. If indications seem favourable and a quiet 'landing' might be reasonably expected, the anchor may be dropped. It may be wise to moor with two anchors or put one out astern, if there is a definite clear patch to sit upon. These areas often shine brightly, even when the rise and fall is great, indicating a useful sandy bottom among the black of rocks or a masking covering of weed. Adjustments may be made by hauling in on the anchors by a later positioning of the kedge or even by the judicious use of a spinnaker pole just before the yacht takes the ground, so avoiding small and until then unseen difficulties. If a long stay is expected, further precautions should be taken, once the yacht has dried out, ensuring that the next low tide is even more comfortable. In the case of small yachts, a certain amount of clearing of the bottom may be undertaken within the swinging circle, so that the possibilities of colliding with an obstacle or a very small rock are removed.

Careful inspection of the chart or the subsequent low tide view may ensure a good position for bird watching. Perching on the top of a sandbank may be good for an overall picture but such exposure may not achieve the close-up that photographers find useful. Reconnoitring is helpful in placing the yacht near to that favoured by a particular species of bird under observation. (See Fig. 15.)

An alternative to the twin-keeled yacht is one with bilge keels. These are extra projections secured at the turn of the bilge and are in addition to the normal shallow draft keel on the yacht's centre line. But there is usually a centre board in this type of yacht to help windward ability, and the main danger when drying out is that the slot may be fouled by small stones and the centre board prove difficult to lower. All in all, therefore, the twin-keeled yacht has the edge when it comes to drying out.

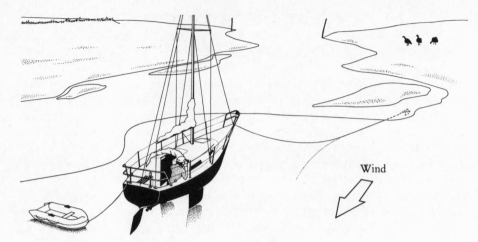

Fig. 15.   Bird watching from a twin-keeled yacht.

However, there is one more form that is designed for this purpose. This is the flat-bottomed barge. The Dutch have developed these over the years to a fine degree. The botter, which is considered the best sea-boat in this flat-bottomed tradition, would make an excellent naturalist's yacht, for as well as acting as a bird-watching hide in challenging places, the vast space below may be well employed looking after the naturalist's other needs. It would not be difficult to fit a small laboratory in such a vessel.

All this does not mean that a yacht with a long, straight keel cannot be dried out; it may be easily adapted to take to ground by the use of legs. The yacht has holes through the topsides and the legs, so that they may be bolted on. Legs are sometimes padded and held tight to the yacht's sides by a single bolt, being secured fore and aft by guys. A pad may be fitted to the foot of each leg if the bottom is soft. The bolt holes are corked when the legs are not in use. A couple of spars may be used as temporary legs by securing tackles to the tops of the spars and leading the lower blocks to the chain plates on either side of the yacht. Obviously the spar must be at least as long as the depth from the safety rail to the bottom of the keel.

Lying Against a Wall

Most experienced cruising yachtsmen have experienced or at least contemplated lying against the quay wall when visiting a small port that dries out, and some of the most picturesque and enjoyable places require this ability. Quays still remaining were often built to supply islands and out-of-the-way places that are now deserted. Vessels that served them were built, as was the Scottish puffer, to rejoice in drying out and, in many instances, without the benefit of a quay. The technique is useful to the yachtsman/naturalist who may find such stone walls a good base for his nature studies.

The yacht is brought alongside, preferably on a rising tide, and secured with long bow and stern warps and springs, so that tidal adjustments are unnecessary. The yacht must be persuaded to lean towards the supporting structure at a gentle angle. Extra weight, perhaps in the form of the anchor chain, should, therefore, be distributed along the side deck nearest to the shore within the run of the keel. There is no use in placing it too far toward the bow for it would be ineffective and could cause, in certain circumstances, the vessel to settle bow down. Anchors, too, can be moved to add weight and increase the list. Another convenient method is to place the dinghy on this side, filling it with water with the deck hose until a recognised angle towards the wall is achieved. This obviates the use of chain, which may be messy (see Fig. 16). Buckets or an empty drum may be similarly employed in place of the dinghy.

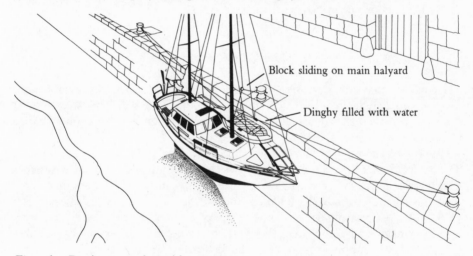

Block sliding on main halyard

Dinghy filled with water

Fig. 16. Drying out alongside a quay.

The yacht's topsides will need fendering and if there are any projections from the quay, a plank suspended outboard of the fenders will be useful.

To add belt and braces, or to be just seamanlike, the main halyard can be useful. Some yachtsmen take the end that is normally secured to the head of the sail ashore and make it fast to a bollard, achieving a list just before the yacht takes the ground. This, however, can twist the sheave and also requires attention at the exact moment of grounding, which may be inconvenient. A better method for the small yacht is to reeve a block onto the halyard, which is then set up taut and to secure this block to a strong point ashore with just enough tension to keep the yacht leaning towards the quay. As the ebb runs away and the yacht falls, the block moves up the halyard maintaining a shoreward pull. This is maintained during the flood.

CHOOSING A DINGHY

The choice of the right dinghy for the yacht has always been a subject for debate. The yachtsman/naturalist has an even greater interest, because much of his study in remote places relies on the efficient employment of the yacht's small boat. Larger vessels engaged on voyages of discovery or on surveying expeditions carried, and still do carry, a large number of boats suited to every purpose.

The alternative to the age-old rigid construction is the modern inflatable, but before looking at the advantages and disadvantages of both types in detail, it is worth examining what the marine naturalist requires. Some of these requirements have already been described under the heading 'Landing in Remote Places'. The boat must be easy to row and to manoeuvre, able to land on rocky shores without damage, able to carry the naturalist's equipment and be convenient to beach and to stow aboard. The dinghy must be able, also, to carry the kedge anchor out in bad weather conditions.

The Rigid Boat

The rigid boat has one overriding advantage over the inflatable. It is usually easier to row and handle in adverse conditions of wind and sea. This statement is qualified because there are some poorly designed rigid tenders that cannot bear to be propelled by oars.

The joy of rowing a well-designed rigid boat should not be lightly dismissed. The ability to place the dinghy exactly where it is required in difficult weather is not the only plus, for there are few more satisfying minutes or hours than those spent gently pulling a well-designed dinghy that will run sweetly after each stroke. This is a pleasure denied to the owner of the rubber boat. The rigid dinghy is usually drier and better able, therefore, to transport equipment that dislikes damp. The conventional dinghy is far better for laying out a kedge or for doing anything that requires rowing into the wind. A rigid boat tows well, too. Finn Engelsen of Bergen, the notable Norwegian yachtsman, employs a long painter, perhaps 30 metres or more in length, and his beautiful varnished dinghy follows his *Anna Marie* like a duckling learning from mother. When stowed aboard, perhaps behind the mast, a small boat may act as a useful stowage, if carried upright and protected by a dinghy cover.

However, this about summarises the list of advantages and the collection of minuses coincides with the plusses of the inflatable.

The Inflatable Boat

The inflatable's main disadvantage, that of being awkward and not pleasing to row, is outweighed by its seaworthiness and high indestructibility on inhospitable rocky shores. It is also light and can be easily carried beyond reach of the sea and may usefully act as a shelter while on land, if secured against the wind. Towing on a very short painter produces little drag unless the boat flips over, which is not unusual. Righting, though, is a matter of moments.

The Avon is one of the best of the small inflatable yacht tenders and comes in eight-, nine-, ten- and twelve-foot sizes. They respond better to a small outboard than to oars, provided that their lack of grip on the water is appreciated by allowing larger turning circles than the rigid keel boat. But with higher-powered engines, the Zodiac wins every time.

Another advantage of the inflatable is that it can be pressed into service as a water carrier. A shore hose alongside is not always available on some of the islands in northern Norway and the technique developed by Agustin Edwards in the cutter *Trauco* in south-western Chile may be used to advantage. The rubber boat is towed by another or swum under a waterfall. It may be bucketed full in a freshwater stream and later pumped aboard, using a mobile submersible, or again bucketed through a funnel into the fresh water tank. It is far quicker than filling and ferrying it in endless plastic water breakers.

A further prime advantage of the rubber boat is the way it will nestle comfortably alongside or lie astern without fear of running damaging forays under the counter or gouging lumps from the transom. In theory, a rubber boat stows well deflated but convenience depends on easy inflation. A properly equipped naturalist's or diver's yacht should have a small air pump or compressor, which makes light work of this. Arthur Beiser uses the exhaust of his vacuum cleaner, for it gives a lot of air fast, finally topping up with the proper foot pump. Using the pump alone is a slow business.

The fabric of an inflatable is tough, but a repair kit and pump should always be available aboard or carried by the shore expedition.

It is clear, therefore, from this examination that for the yachtsman/naturalist, the inflatable is the best choice, if only one boat is carried. However, to combine the best of both worlds, the answer is to carry both an inflatable and a rigid boat. The rubber dinghy may be stowed inside the conventional one and this should find a place just aft of the mast on most sailing yachts. The rigid boat will need extra buoyancy under the thwarts for rough water work, and a small compass set in the side thwart parallel to the centre line is a good idea, too. If there are no side thwarts, a protected 'binnacle' built onto the floorboard should be in the view of the oarsman. A pocket compass may well be forgotten and so such a fixture is useful. It is a good way of keeping a straight course, even on a fine day, when rowing.

Rowlocks are important too. At present, plastic of any sort should be avoided. They appear firm and efficient but give way under strain. Captive metal rowlocks are the answer, either secured by rings to the oars themselves or folding up from the gunwhale. The former have the edge, as the idea is simpler, and with this method, even spare oars are armed with their own rowlocks. The Zodiac system of pins in the inflatable is to be preferred to the Avon's rubber rowlocks, provided that the boat is properly inflated and remains so. The Avon has an awkward feel about it because of the rubber rowlocks, neat though they appear at first sight.

Bailers are important in all boats and are better than a pump as they cannot go wrong and may be employed for other purposes. The plastic dustpan type can reach almost to the last drop, and is useful for removing grit and sharp shingle which can do damage, particularly to inflatables.

A dinghy anchor, perhaps a folding Norwegian grapnel, should be made up with its own line spliced on and kept in a convenient box for placing aboard the dinghy. An inordinately long painter, doubling as an anchor warp, is a nuisance when used for ordinary purposes. An anchor bag offers a solution, for it contains both the anchor and warp, as well as a spare length of line for places with a large tidal difference. A block should be included so that extra yardage can be used with the anchor or a convenient stone as a dinghy outhaul from the beach. The bag ensures that everything is together and easy to stow, either aboard or in the dinghy itself. Finally, on the matter of warps, a stern line of six metres or so should be spliced to the dinghy, as this is useful for securing alongside or for hoisting the boat itself aboard.

The skeg of the rigid dinghy needs to be protected, as do the bilge rubbers, with a brass strip, and a permanent, thick, rubber fender all the way round is essential. There are few more useless things than the baby dinghy fender.

The final point on the rigid boat itself is construction. Should it be wood, carvel or clinker-built or cold-moulded, or should traditional materials be abandoned in favour of plastic or aluminium? There is no doubt that the best-looking of the lot is the clinker, double-ender beloved of the Clyde. The Americans, too, have some fine wooden dinghies, such as the classic pulling boats of L. Francis Herreshoff, still occasionally built in the small boat workshops of Mystic Harbour. There is nothing like a teak dinghy with a rock elm grating aft for looks or for demanding well-deserved care and attention. It would be a shame though to let the yachtsman/naturalist loose in one of these, for if he is enjoying his work pursuit properly, his dinghy would show signs of it on every strake.

A cold-moulded dinghy has great advantages, for it looks good and, as Hal Roth has said, extras may be screwed to it. However, fibreglass now has the edge because so many are made and the choice is legion, from the speedy small Wasa range to the simulated clinker. The boat, which should be the largest the owner can conveniently handle, must fit aboard so that the boom can pass over it, whichever way up is chosen, and it should not obscure the view forward.

CHAPTER FIVE

The Naturalist's Yacht and Equipment

CHOOSING A YACHT

The sailor/naturalist may use any type of craft in his exploration of the natural world, from an inflatable dinghy to an out and out ocean racer. Even the fast power boat forms a platform which may be used with care. However, if this exploration becomes one of the principal pleasures of sailing, there are a number of points which should be borne in mind in selecting or building a yacht for the purpose.

The first is that she should be both at home and a home at sea. The naturalist's yacht must be seaworthy and a stable platform for observation and work. Generally speaking, this means that the light, high performance yachts requiring constant attention to obtain the best of them are not ideal. Observation of the natural world is, after all, an additional and time-consuming task at sea and both hull and rig need to reflect this so that some of the time that is normally devoted to the sailing of the yacht may be diverted to observing the life that surrounds the vessel. A vessel designed for single-handed sailing but now crewed with more than one person aboard, may be ideal, particularly if that yacht is also sea kindly and stable. The fallacy that cruising boats are born of pensioned-off ocean racers is widely held. The design of a good cruising boat is a difficult and demanding task and requires as much skill as that required to produce a boat that can win ocean races. However, a really successful cruising yacht may well make an excellent exploration vessel.

The second requirement of the naturalist is that his yacht is 'handy', that she may be navigated with confidence into small corners and will lie comfortably at anchor in an open roadstead. A further advantage allied to the above is covered in Chapter Four. This is the ability to dry out either against a wall, on legs or twin keels. This enables the sailor to explore areas denied to those rolling along on tender skegs, exposed rudders or screws.

The third point is the need for good stowage space. The sailor/naturalist carries more equipment than the ordinary yachtsman. Some of his equipment is fragile, requiring built-in protection. His shore-going gear, too, is specialised. Some equipment, such as the livewell and microscope lamp, have peculiar demands on the yacht's electrics. If the naturalist is also a diver, he will require a place for air bottles, perhaps a compressor, and a special wide boarding ladder that will take fins.

PENN FATHOM M

LOC
SI

⑦

WC
UNDER
SEAT

⑩

SHELF

COCKF

CHARTROOM

⑥

CHARTS

⑨

⑩

③

LOCKER

①

⑧

⑦　⑥

⑩

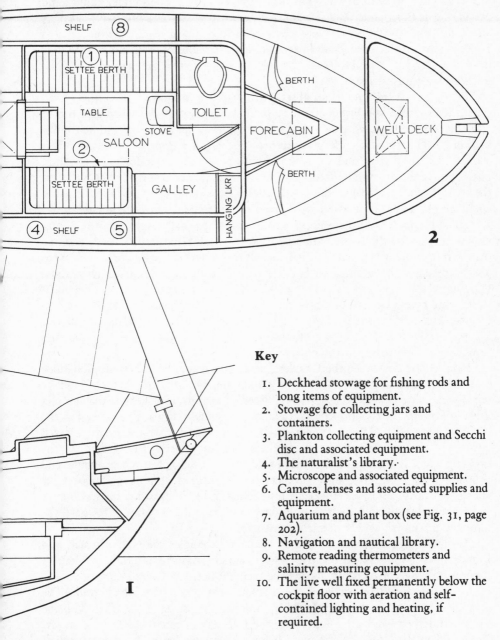

OWN RIGGER FOR PLANKTON NET AND FISHING

SHELF ⑧

① SETTEE BERTH

TABLE

SALOON

② STOVE TOILET

② SETTEE BERTH

GALLEY

HANGING LKR

④ SHELF ⑤

BERTH

FORECABIN

BERTH

WELL DECK

2

I

Key

1. Deckhead stowage for fishing rods and long items of equipment.
2. Stowage for collecting jars and containers.
3. Plankton collecting equipment and Secchi disc and associated equipment.
4. The naturalist's library.·
5. Microscope and associated equipment.
6. Camera, lenses and associated supplies and equipment.
7. Aquarium and plant box (see Fig. 31, page 202).
8. Navigation and nautical library.
9. Remote reading thermometers and salinity measuring equipment.
10. The live well fixed permanently below the cockpit floor with aeration and self-contained lighting and heating, if required.

Fig. 17.　A Darwin class expedition vessel, profile (1) and plan (2).

The need for such equipment, some of it necessarily heavy, spells out the need for a load-carrying hull. A sailing catamaran provides an excellent stable platform and will dry out well, but may not be handy enough under canvas and the sailing performance of multi-hulls suffers if they are heavily overloaded. A small, all-power cat, in relatively sheltered waters, has great attractions, as it is an ideal platform for dredges and nets. Similarly, the Boston whaler or Dell Quay dory are excellent trailable boats for the naturalist. They are stable, can carry loads and are unsinkable.

The ideal craft for the serious observer though is a full-powered sailing vessel. The first example, the Fisher 25, may be more accurately described as a motor sailor. It is difficult to imagine a 25-footer with more space aboard and her hull is sea kindly. Her rig and underwater profile enable the engine to be turned off, if speed is not important, and yet is there, with power to spare, for manoeuvring, positioning and for towing dredges and lines at the right speed without wear on the crew. Both the wheelhouse and the saloon have room for study and the yacht could be fitted out so that all the necessary gear has a convenient and easily accessible home. The Fisher adopts a hull similar to that developed over a long period of time by the fisherman, who, like the sailor/naturalist, needs a stable platform to work gear, whether it be shellfish dredges or nets. Such craft were also accustomed to use small harbours of refuge at the foot of cliffs, such as those clinging to the shores of the south-eastern shore of Moray Firth, on Scotland's north-east coast, or harbour protection grafted onto the rocks, as in western France. The need for sea kindliness and the occasional necessity to dry out dictate a long, straight keel. The wheelhouse on the Fisher extends the season and provides further shelter in the cockpit.

Colin Mudie developed the Charles Darwin class for the waters inside Chiloe and the Chonos archipelago in southern Chile, but this 28-footer would be equally at home in the Norwegian or Swedish islands and skerries that protect much of their rockbound coasts, which are such a rewarding area for the seaborne naturalist. The arrangement shown in Fig. 17 is for a small research vessel and the stowage of the naturalist's equipment is clearly shown on the opposite page. A feature of the little exploration vessel is the wheel shelter and charthouse aft. This not only protects the helmsman and navigator and his charts, but provides a peaceful place for the naturalist to examine his specimens. In harbour, this unusual addition does duty as a loo at the 'bottom of the garden', a luxury unique in yachts of such small size.

The down rigger mounted on the cockpit rail handles the plankton net, and fishing lines and a Neco electric power winch may be provided to haul a naturalist dredge or to bring aboard the triangle net. On *Gang Warily II*, a 30-foot cutter owned by one of the authors, this winch worked well. The arrangement, was further developed by hinging the aft davits, so that with the yacht's dinghy towed on a long line, the two davits acted together as an 'A' frame, when the 'stop' on the warp before the dredge lifted the frame from its position over the stern and brought it into a near vertical position. This meant

Plate IX

SIPHONOPHORA AND JELLYFISH

1. *Muggiaea atlantica* : Siphonophore
2. *Sarsia eximia* : Medusa
3. *Physalia physalis* : Portuguese man-o'-war
4. *Velella velella* : By-the-wind sailor
5. *Cyanea lamarckii*
6. *Chrysaora hysoscella* : Compass jellyfish
7. *Aurelia aurita* : Common jellyfish
8. *Rhizostoma pulmo*

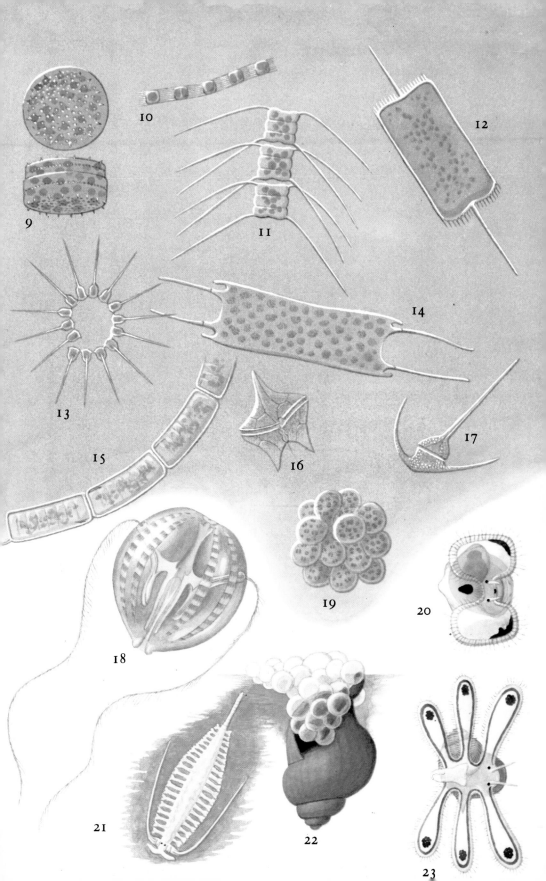

9

10

11

12

13

14

15

16

17

18

19

20

21

22

23

Plate IX

PHYTOPLANKTON AND ZOOPLANKTON

9. *Thalassiosira excentricus*
10. *Skeletonema costatum*
11. *Chaetoceros densum*
12. *Ditylum brightwellii*
13. *Asterionella japonica*
14. *Biddulphia sinensis*
15. *Rhizoselenia alata*
16. *Protoperidinium depressum*
17. *Ceratium tripos*
18. *Pleurobrachia pileus*: Sea-gooseberry
19. *Phaeocystis pouchetii*
20. *Littorina littorea*: Veliger larva of edible periwinkle
21. *Tomopteris helgolandica*
22. *Janthina janthina*: Violet pelagic snail
23. *Aporrhais pespelicani*: Veliger larva of pelican's foot shell

Plate X

ZOOPLANKTON

1. *Mytilus edulis*: Veliger larva of common mussel
2. *Lepas anatifera*: Goose barnacle
3. *Balanus balanoides*: (a) 5th nauplius and (b) cypris of acorn barnacle
4. *Calanus finmarchicus*
5. *Evadne nordmanni*
6. *Acartia longiremis*
7. *Leptomysis gracilis*
8. *Centropages typicus*
9. *Euchaeta norvegica*
10. *Temora longicornis*
11. *Mesopedopsis slabberi*
12. *Meganyctiphanes norvegica*
13. *Praunus flexuosus*
14. *Parathemisto compressa*
15. *Eupagurus bernhardus*: Megalopa of hermit crab
16. *Cancer pagurus*: Megalopa of edible crab
17. *Carcinus maenas*: (a) 3rd zoea and (b) megalopa of shore crab
18. *Palinurus vulgaris*: Phyllosoma of spiny lobster
19. *Crangon vulgaris*: 1st larva of common shrimp
20. *Porcellana platycheles*: Zoea of hairy porcelain crab
21. *Macropipus puber*: 4th zoea of fiddler or velvet swimming crab

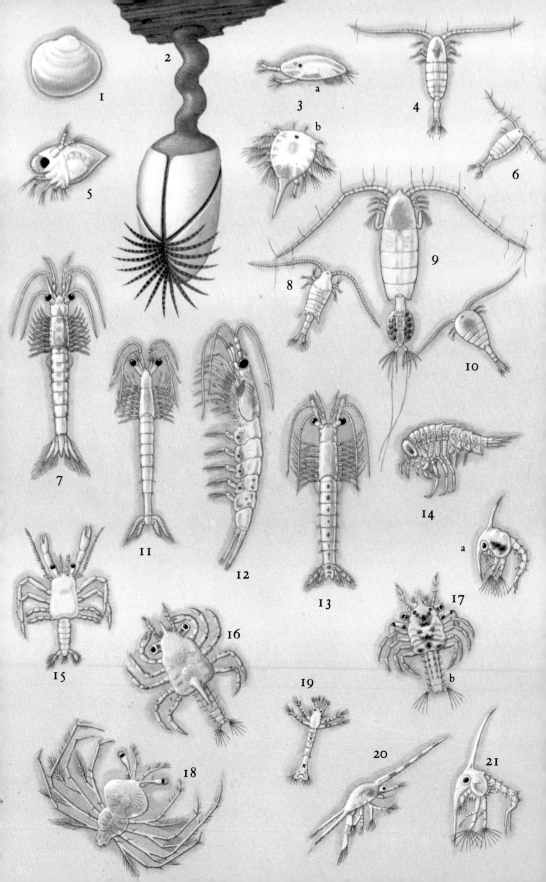

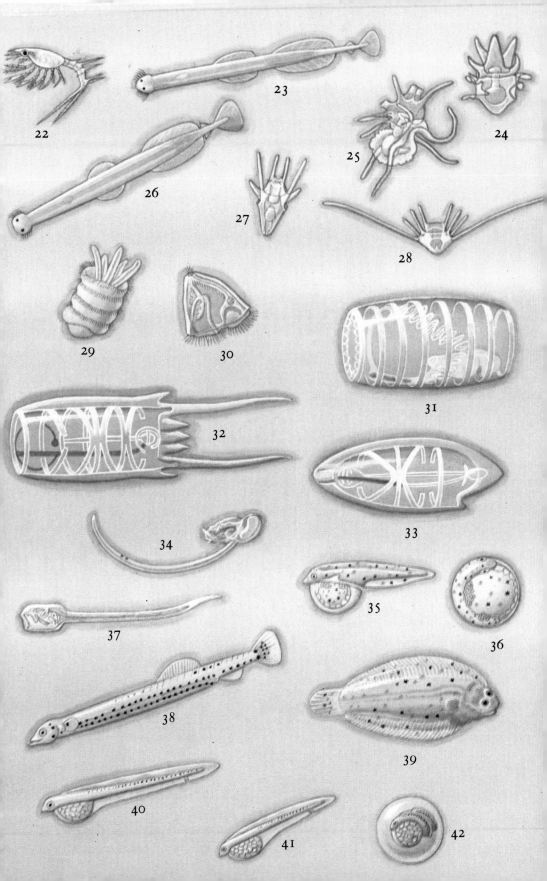

Plate X

ZOOPLANKTON

22. *Nephrops norvegicus*: 1st zoea of scampi, Norway lobster or Dublin Bay prawn
23. *Sagitta elegans*: Arrow worm
24. *Asterias rubens*: Bipinnaria of common starfish
25. *Asteria rubens*: Brachiolaria of common starfish
26. *Sagitta setosa*: Arrow worm
27. *Psammechinus miliaris*: Echinopluteus of green sea-urchin
28. *Ophiothrix fragilis*: Ophiopluteus of brittle star
29. *Holothuria* sp: Pentacula of sea-cucumber
30. *Membranipora membranacea*: Cyphonautes of sea-mat
31. *Doliolum geganbauri*
32. *Salpa democratica*: Solitary forms
33. *Salpa democratica*: Aggregate forms
34. *Oikopleura dioica*
35. *Solea solea*: Newly hatched larva of sole
36. *Solea solea*: Egg of sole with embryo
37. *Clavelina lepadiformis*: Ascidian larval form
38. *Clupea harengus*: Post-larval stage of herring
39. *Solea solea*: Post-larval stage of sole
40. *Clupea harengus*: Herring larva
41. *Sardina pilchardus*: Pilchard larva
42. *Sardina pilchardus*: Pilchard egg with embryo

that the contents of the dredge could be emptied on the open-ended deck aft, above the transom, for sorting. Such an arrangement is not possible on the Charles Darwin class because of the 'caboose' and the dredge or triangle net must be handled over the side.

The yacht's saloon has a coal- and wood-burning stove. Driftwood, gathered dry from the top of the strand line, makes an attractive, free fuel.

Note the straight keel and the sculptured stern post. The rudder blade is able to rise, too, and so there is no danger of a bent pintel, if the vessel is beached or awkwardly launched. The Charles Darwin class is designed to be trailed, so widening the horizon and cruising ground of the owner with limited time.

The Fisher 25 and the Colin Mudie yachts are but two examples, the former a standard design well suited and the latter thought out for the purpose. It is worth emphasising again, though, that any seaworthy craft that will allow her crew time to look around safely in reasonable weather, is capable of being transformed into a useful exploration vessel for the naturalist. Interest, ingenuity, seamanship and a keen eye will achieve the rest.

THE EQUIPMENT AND HOW TO USE IT

As the sailor/naturalist becomes keener and more aware of his surroundings, he will start to collect equipment to aid his study of the marine environment. He will devise his own gadgets and gilhicks which will help him observe, record and stow his findings and catch.

The Yachtsman's Journal

Recording 'day to day' happenings in the form of a notebook or log is an essential part of research and observation. The record is, perhaps, the most important part of observation, for if experience is to be written down, the eye becomes sharper and more discerning. It would be little exaggeration to say that the whole purpose of the log is to produce a readable and accurate record of the marine environment. This may seem one of the labours of Hercules, but record- or log-keeping should not be done as a chore—the end result must be decorative as well as packed with information. Many yachtsmen already keep a log—it is a seamanlike thing to do—and there are many standard logs or bound collections of printed forms to make this easy. They are, however, as convenient as they are dull and the majority of such records find their way to the tip or wastepaper works when a yachtsman 'swallows the anchor', or passes them on to an honoured relative. It is a shame to lose any record, but it is usually the recorder's fault if he fails to capture the interest of others by not broadening the appeal.

The yachtsman/naturalist has many opportunities to make his observations both useful and decorative. The page margin, for example, can be used to describe a walk down the shore, showing the zonation of species (see Chapter One, Fig. 2). Individual species of plants and fishes are well worth recording, both by photograph and by sketch, pointing out in the text the various points of interest.

Sketching and Illustrating

Sketching is seen by some as a black art or an inherited gift and yet most people, given an ounce of courage and by filtering self-criticism, can achieve a 'record'. All that is needed is a set of drawing pencils (HB), avoiding using those reserved for chart work, a sharp knife (essential to the sailor in any event) and a soft rubber. Again, there must be the same discrimination against using the navigator's tools of the trade or they will not be either there or in proper condition when he really needs them.

The midshipman of yesterday was taught to sketch and record views and profiles of the land from the seaward. The editors of sailing directions today employ, for the most part, photographs.

There is a profusion of small harbours, creeks, inlets and remote landing places that go unnoticed and unsung in a growing collection of sailing directions. Many of these will be of particular interest to the sailor/naturalist and his log or journal should be a good resting place for such information, whether in note or note and sketch form. The photograph, too, has its place. The Royal Cruising Club encourages members to contribute their information and findings to a series of folders devoted to coasts and harbours all over the world. This 'Foreign Port Information' is of great assistance to others in the Club. The record sheets are laid out in a set way which helps the would-be contributor as well as those who use the information. The heading highlights the 'Name of the port', 'Latitude and longitude', 'Member who contributed the information', 'His yacht and date of survey'. This is followed by paragraphs including a brief 'Description of the harbour', 'The approach', 'Anchorage', 'Formalities', 'Facilities and remarks'.

The value of such information in the log or journal is much increased by a sketch plan and view, as shown in Fig. 18. Note the wealth of information that is included in both the plan and the profile. The sketch of the lighthouse provides a description useful for further visits or for others who might use the notes later. The same is true of the record of the entrance, the anchoring places, the small building and pier and the general lie of the land. The sketch profile of the hills behind is an important aid to recognition too, particularly in times of bad light or on a moonlit night.

The naturalist, without much of the artist in him, can also enhance his records by simple sketches. A start can be made by sketching a shell from the beach, using as an example one of the identification manuals that employ drawings rather than photographs. It may be a help to trace round the half shell of a bivalve, placing it hinge downward to the paper. Later, those points that distinguish them from their cousins can be shaded in. Obvious details of

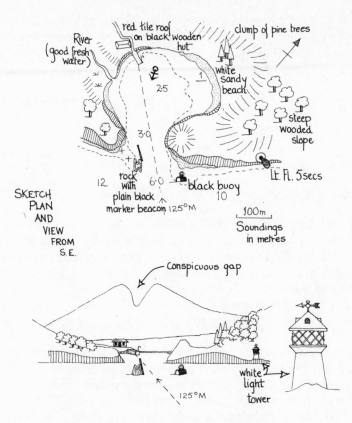

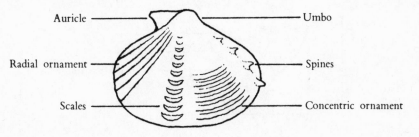

Fig. 18. An imaginary harbour sketch, plan and view.

difference are the 'growth stages' in the form of rings and the radiating ribs (see Fig. 19).

Recording the measurement of a shell is important, too. Bivalves are measured from one end to the other when the hinge, external ligament and beak are uppermost. The edges of the shell are called the anterior and posterior margins.

Fig. 19. External features of a bivalve.

Auricle ———————————— Umbo

Radial ornament ———————————— Spines

Scales ———————————— Concentric ornament

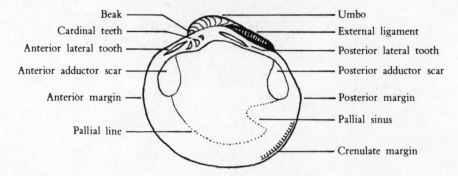

Fig. 20.   Parts of a bivalve: interior view of left valve.

Even the strange shape of a razor shell is treated in the same way and so with the external ligament upward, the margins are well separated. There are points of difference on the inside, too, and another view can be shown, emphasising the pallial line and adductor scars. The cardinal teeth are often important in identification and may be sketched as an inset (see Fig. 20).

A gastropod has a different set of dimensions. The width is, as you would expect, across the shell and the height from apex to siphonal canal at the other end (see Fig. 21).

The measurement recorded on the roughest of sketches gives an added dimension, as do notes with arrows. A page reserved for recording the shells of a particular length of coast can add to the pleasure of collection and allow sketching to be done when awaiting a fair wind or weather.

Plants, fish and birds can be treated in the same way. Notes alongside make up for any deficiencies in the artist. Direct comparisons 'like a robin or a wren' may

Fig. 21.   External features of a gastropod.

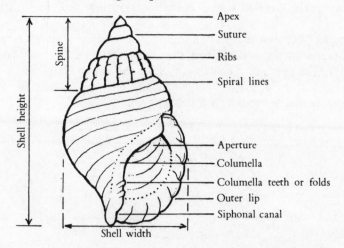

not be particularly scientific but mark the writer's memory and give an impression to others. Plants can be treated in a similar way.

The paint box can add further dimension. Small ones that are easily stowed are available, some with their own fresh water container on one end. This advantage is not essential on board, as a container with a lip will do, but it has advantages on the beach. A selection of brushes is required and in order to fit them into the box, it may be necessary to trim the wooden ends. It is also best to protect the brush points by putting a short length of plastic tube over that end, so that the tip is not continually flattened as the ship rolls. Pencils may be similarly protected.

All this material is best stowed in a clearly marked box. As the artist has been forewarned about robbing the navigator, the navigator must be similarly advised—he must not rob the artist. If they are one and the same person, he should keep his equipment separate.

Felt-tip pens improve the writing and have their uses aboard. The trouble is that they are not usually waterproof and if they are, the spirit containing the ink goes right through the page, spoiling the reverse. If felt-tip pen sketches are decorated with water colour, particular care must be taken to avoid smudging. The ink, when watered, goes a peculiar brown colour which can, however, be turned to good effect, using the smudge, as it were, for shading, by distressing the edge of the pen line. Black or brown water colour can be added by brush, thus lending further tone to the result.

Photography

The photograph is nearly as important as the written word and can be as effective as a good drawing. Further, it can enhance the journal, bring to life not only the experience but the people who shared it, and act as a tool of discovery.

There are some good books on nature photography in the bibliography. The section here is not designed to teach the sailor how to use a camera—it assumes he can 'snap'. The hints are designed to help capture and record what he sees without going into complicated technical explanations.

The simplest camera to use is the 'instamatic'. It is not a close-up camera but it can be useful in capturing the beach scene and can usually focus within a metre of the subject. The most effective use of this camera is to employ it for general views and natural history portrait work, remembering that detail will not come easily. This is where the photographer/artist comes in, for detail can be added with pencil, pen and water colour, referring to the photograph.

In the portraits, it is necessary to introduce scale and this is usually done by including in the photograph a familiar object of known length, such as a sailor's knife or a geologist's hammer. The lanyard of the former can be made more useful by making it an exact metre overall, or marking it off accurately with a waterproof felt-tip pen, so that it becomes a measuring tape. Height and distances then may be recorded easily in the field.

The pocket notebook should be another constant companion. This becomes the 'true' rough log, leaving the 'smooth' edition dry on the chart table. Each scene is noted against the exposure or picture number; in this way:

'Film 2—6 (exposure number) Lepe Beach—Solent (south of England) strand line—3/4 ebb—weak sun—plastic cups, driftwood and 3-stranded "4-eyed" jellyfish—120 mm. across—*Aurelia aurita*?'

The film number is useful and the exposed film storage container should be so marked for later reference. The number of the exposure and place gives a clue when the picture is stuck in. The position can be pencilled in the log in advance, perhaps writing the caption from the notes, when back on board and before the photograph is developed.

The final paragraph in the log developed from the note above might read:

'1445: We anchored at the first bend in the entrance to the Beaulieu River. George had noticed "mammal bones", marked on an old geological survey map and we went ashore to investigate. The tide was nearly "all away" and the shingle, sand and mud of the lower shore was exposing fast. The beach was pocked from the efforts of bait-diggers: what a mess they make of the beach—several of them were at work chasing the lugworm, *Arenicola marina*. No chance of finding mammal bones here now, the place looks more like the Somme after the battle. We wandered up the strand line of the shingle beach to do a little "combing". There were more of those "4-eyed" jellyfish *Aurelia aurita*. We had seen plenty on the surface of the Solent on the way from Cowes, Isle of Wight. There was not much else exciting though, as one can see from the photograph. We were back on board at 1600 and weighed soon after, making for Bucklers Hard under power, as the wind had died.'

Note the name of the jellyfish is now positive and was confirmed from a field guide drawn from the yacht's bookshelf.

All types of cameras can be adapted to close-up work with the aid of supplementary lenses. These may be used in combination, so bringing the subject nearer in defining detail. They are simple to use and instructions are given with the lenses. It is necessary to measure accurately, however, the distance between the lens and the subject when not using a single lens reflex camera (SLR), which enables the user to see the image through the lens.

Extension tubes may be fitted to cameras with a removable lens but bellows provide a more sophisticated way of achieving closer work, as well as an infinite range of focus. They are more expensive than the other alternatives. Further techniques, such as reversing the lens on the tube, or bellows, with a reversing ring, or coupling two lenses together, are available to the expert. The important thing to remember with bellows or tubes is that the exposure has to be increased to avoid under exposure. Tables are included though with the equipment and this takes the guess out of their use.

Easily the most convenient way of taking close-ups is the use of a macro-lens on an SLR camera. This gives positive focus and a flexible choice of speed and aperture at the fingertips.

A 35-mm Nikkormat camera body with a micro-Nikkor-P Auto 1 : 35 f = 55-mm lens will take all the pictures the usual yachtsman/naturalist could wish for. Other makers have their equivalents, equally as effective for the purpose and although perhaps less well made, at less cost. The 'general scene' and a 'close-up' may, therefore, be recorded with one camera and sketches become less important, though always an asset.

In photographing close-ups, the maximum depth of field may be important or only part of the object will be in focus. In simple terms this means decreasing the size of the aperture by using a higher 'f' number and reducing the shutter speed. There is usually a scale on the lens that helps in this calculation or a button that shows what is in focus and what is not, before the picture is taken.

A few words about lighting when capturing beach flora and fauna. Most yachtsmen/naturalists will make best use of available sunlight and not use a flash or other artificial light. Additional light is useful though for picking out individual flowers from their habitat, at the same time increasing the depth of field on the flower itself. It also enables a higher shutter speed, which is helpful in windy conditions. Screens, too, can be brought in to help to still the gyrating flower head, a problem familiar to both amateur and professional photographers.

The angle of the picture is obviously important and the identification of the species is easier against the sky or, indeed, against water. Occasionally it is possible to capture the subject in bright sun while the background is in shadow, but this is a short-lived opportunity.

Flash is also useful for fauna under rocks and a battery-powered, hand-held electronic model is preferable to the rechargeable alternative when kept aboard ship.

To achieve slow speeds and difficult angles, a tripod is essential. This should be small and so provide a stable platform at a low height. Something around a foot or 30 cm is ideal. A spike is also useful and may be half the length of the tripod with the usual ball and socket head to mount the camera. The spike itself can be made at home, or even on board, from wood or metal, provided the top is available. It is obviously essential to have a cable release to operate either of these to avoid camera shake.

Filters, as their name implies, change the make-up of light reaching the lens and, therefore, the film. There are three filters that should find a home aboard the naturalist's yacht. The contrast filter, for devotees of black and white, will record the difference between similar shades of red or green. The sky or haze filter is especially useful for boats, as it overcomes the blue tint in colour films exposed to a great expanse of sea and sky. The last and the most expensive of these desirable aids is the polaroid filter. This is invaluable for probing shallows or sides of the ship's aquarium. The filter does this by cutting down the polarised light reflected from the glass of the aquarium or the surface of the water. It is useful for both colour and black and white film. By twisting the filter itself, the effect can be varied and this is easily seen through the viewfinder of a single lens reflex camera. The exposure should be increased by a couple of stops to compensate for the loss of light.

To be closer to the subject when examining creatures in shallow water, the camera lens needs to be below the surface. There are sophisticated techniques for doing this, using a periscope, but these are expensive and complicated. A glass bottom box achieves the same end both for photography and for observation. The camera should be mounted on a frame on the top of a box, protected by a flap with a spy hole, with an exit for the shutter release cable. This ensures that spray cannot attack the camera and that reflections do not mar the view returning from the glass window below the lens. The box should have provision for weights (or it will be difficult to control when partially submerged) and sufficient room for an electronic flash.

With the camera now in the water, it is worth considering what to do if the box floods or the camera drowns. Insurance is one answer but a more merciful one is to wash the camera immediately in fresh water, followed quickly afterwards with a bath in methylated spirits or alcohol.

Photographing With An Aquarium

The shipboard aquarium, described on p. 196, is a useful way of capturing what goes on in the sea either by observation or photograph. Background is important and the small aquarium suggested for shipboard use has provision for a variety of coloured backgrounds that can be slipped in easily. The darker the colour the more obvious the bubbles adhering to the front glass. A water colour brush can be used to remove these before a photograph is taken. Reflection again is to be avoided and a good way to overcome this is with the use of a black card mask and by positioning the flash above the tank at 45 degrees to the water.

The sun shines through the surface of the sea illuminating the animals and plants below. Overhead lighting of the tank is, therefore, the most natural. The chart table may well be the best place for shipboard aquarium photography and a flexible chart light may be used to help position the flash, as experiments with shadow should be performed before pressing the button. This means, though, that a spare, powerful chart light bulb should be kept. This could also be useful for microscopy, as is explained later.

Underwater Photography

Underwater photography should not be attempted before one is familiar with use of the camera on land. The principles of taking pictures underwater are much the same as on the surface, but the conditions are almost always worse than those which would be considered totally unsuitable for photography on dry land. Visibility by comparison with surface conditions is always poor and it is not worth trying to photograph objects which are more than a third of the distance from the lens than the visibility range. Close-up and macrophotography are well worth considering under water, as good results can usually be obtained in all but the worst visibility. Light levels are generally low and colours disappear very quickly with depth; surprisingly, film often fails to pick up colours which are still apparent to the diver and so flash is recommended for most work. Electronic flash units are more expensive to buy initially than bulb units and they generally produce a lower output of light. They are, however, far more convenient to use because spare bulbs do not have to be carried by the diver and the cost per shot is considerably less.

Colour transparencies and colour or black and white prints can be taken with equal success under water, but on the whole colour is to be preferred, particularly if a flash is being used, as many of the shades and tints of the underwater world are too good to be missed. The speed of film chosen will depend on the lighting conditions anticipated, the power of the flash and the distances from which subjects are to be photographed. In general, slower films give crisper pictures and, where colour is being used, more brilliant results.

Underwater cases for surface cameras can be constructed by the handyman with the right tools but access to workshop facilities makes the task much easier. The material usually chosen for building cases is perspex, which can be bonded to give watertight joints with a proprietary cement; controls may be obtained from dealers who advertise in diving magazines, and ingenious amateur engineers have devised all sorts of lever and pulley systems to operate shutter release, film transport, focus, shutter speed and aperture controls. Details of design and construction methods are given in Horace Dobbs's excellent book which is listed in the Bibliography.

Housings can also be bought ready-made and they range from simple perspex cases to the elaborate and expensive Rolleimarine housing for the Rolleiflex twin lens reflex cameras. Probably the best answer for underwater photography at present, however, is the Nikonos underwater camera which is produced by Nippon Kogaku K.K. of Japan, the company who also make the excellent Nikon range of cameras for use on the surface. This camera was first designed for Cousteau by the Belgian engineer Jean de Wouters and manufactured in France as the Calypso-phot. Production was later taken over by the Japanese company and the camera given the name 'Nikonos'. The latest version, at the time of writing, is the Nikonos III. It is no larger than a conventional 35-mm camera and consists of three basic units, the largest of which incorporates the film transport system and a focal plane shutter; this seals into a pressure-resistant aluminium shell with an o-ring, and the lens assembly with aperture and focus controls locks the three units together. A bulb flash unit and a variety of different lenses are available from Nikon and a wide range of accessories including electronic flash units and extension tubes for close-up work are made for use with the Nikonos by other manufacturers. One of the great advantages of the Nikonos for yachtsmen is that it doubles as an excellent surface camera which is completely resistant to salt, spray, sand and dust and is rather more robust than most cameras built for surface use.

Photographing Insects

Although insects do not feature in the descriptive section, they occasionally arrive aboard, long distances from land. A collecting net at the masthead could produce some interesting results and a pooter and a larval breeding cage enable the catch to be photographed at sea. A pooter is a small mouth-powered vacuum cleaner for selecting insects. Breeding cages are now supplied in high-impact polystyrene with a clear plastic cylinder and so are rustproof and suitable for shipboard use. Another interesting piece of equipment, serving as a collecting net aloft, is the sticky sail, which works like a sophisticated flypaper.

Stalking insects with a camera is best done with a single lens reflex and a telephoto lens. Good pictures can be taken with close-up equipment by those with patience. It is best to set the focus on a predetermined distance and then approach carefully until the prey is in sharp relief. A halting stalk followed by fumbling with the lens is likely to fail.

Photographing Birds

Perhaps one of the most challenging creatures to photograph are seabirds. The telephoto or long focus lens is essential for the bird photographer afloat. If any convincing is needed, the results of trying to capture on film a bird in flight with a normal 55-mm lens should do the trick. Lenses from 125 mm upwards to 300 mm, until they become too heavy to hold steady, give excellent results.

The yachtsman has some advantages over his land-bound colleagues. A small boat, properly handed, can act as a hide. A certain amount of cunning on how a camera, as opposed to a gun, may be used to advantage in estuaries, may be gathered from the works of such famous English wildfowlers as Sir Ralph Payne-Gallwey and from Colonel Peter Hawker. They often used sail to bring them within punt gun range of rafts of duck, ensuring that the rising or setting sun was behind them, making their approach difficult to spot.

The yacht has other advantages, for it is a hide in itself. Careful anchoring, as has been discussed, with the use of legs or twin keels, can place a vessel in spots that waders frequent, so putting the photographer among his quarry. It is also an advantage to land at low water to set up a hide on the shore and allow the incoming tide to slowly drive the birds towards the lens.

Drying out in an estuary requires care. At low water the illusion of shelter may be misleading while at high tide problems may develop. Such matters have been dealt with in the chapter on seamanship. The possibilities, with proper planning, are legion. The sea and sands behind the Dutch and German Friesian islands, that begin the great productive region of the Wadden Zee, are of intense interest to those interested in seabirds. Texel (pronounced 'Tessel') is a place of particular fascination. Texel Island starts the chain that reaches east into Germany and includes the Dutch islands of Schiermonnikoog and Rottumeroog. In the spring and autumn the waders come in their thousands. No wonder the place captured Erskine Childers' imagination—he set his classic sailing tale *Riddle of the Sands* in the waters round Borkum and Norderney. *Dulcibella*, his yacht, was at home there and an ideal craft with her ditch-creeping draft of less than a metre.

Microphotography

Most single lens reflex cameras have adapters suitable for standard microscope tubes. Polaroid make a special camera for the purpose, so allowing instant results. Trial and error is perhaps the surest way of obtaining a satisfactory record. Each exposure should be recorded, making notes of exposure number, time and the f. stop, so that lessons can be learnt when the results come in. There is less waiting with the Polaroid, though the film, for such trial and error, is expensive.

Sound Recording at Sea

Sound recording at sea is a new thought but with directional microphones and sophisticated accessories that filter out background noise, the recorder of seabird calls or for that matter of any sea noises, has a new field to himself.

The great seabird colonies of Marwick Head and St Kilda, the waders of the Wadden Zee, or the black-headed gulls of the Solent are ready for the tape. The yacht can be as efficient a hide for the recorder as it is for the camera. It is possible to listen to and record the noises of animals below the keel by using special microphones designed to catch underwater sounds.

Labelling Specimens

So far concern has been confined to the log and logging but this is not the whole story: specimens may be collected and they require labelling. Plants collected for the flower press, described later, require notes, preferably in a standard form so that the final collection has meaning. The example down in Fig. 22 comes from Kew and shows the proper length of observation. The yachtsman/naturalist may content himself with the same headings, but use less detail. Plain sticky labels are available that will suit almost any amount of record. The printed form, too, can be dispensed with, using a waterproof pen to supply the headings.

Oceanographic Measurements

The methods of recording described so far are enough for most people, and the simple research ideas given below may take over too many useful hours that might be well employed enjoying the helm or the set of the sail. Others, though, may wish to go further and look at the anchorage or chosen area in more detail. Each of these small investigations can add interest and there is an infinite variety of opportunity.

BRITISH WEED FLORA

Scientific Name: *Plantago coronopus* L. (*Plantaginaceae*)

Common Name: Buck's-horn Plantain.

Locality: Trimingham, Norfolk - on the coast north-east of Norwich. West of the A148.

Situation: Gravel pit, sandy soil and damp in wet weather.

Frequency: Very common.

Notes: Stamens pale yellow. Fr. 3-4 seeds dehiscent. It is common on sandy soils and is frequently found on golf courses and bowling greens, especially near the sea.

Date: 25.9.76. Annual/Perennial. Fl. May-Sep.

Collector: E.A. Leche. No. 54.

Fig. 22. Label of British weed flora. This illustrates the correct way of labelling herbarium material. The important points are:
(i)   identification & nomenclature
(ii)   preparation (pressing & mounting)
(iii)   completeness of material
Kew Diploma students, for example, are expected to provide: seed, cotyledon stage, seedling stages, mature plant with root, flower and fruit. This will be difficult for the yachtsman unless he keeps returning to the same location, usually impossible on expeditions and voyages.

As will be seen in the previous chapters, the kind of plant or animal found in any area depends upon the seawater environment. Some species are more tolerant to change than others and can be found in both sea and estuary. Simple physical measurements demonstrate the differences between estuarine and sea waters and these may be looked at with the results of biological observation. The animals and plants that are unveiled as the tide recedes give a good clue to the composition of the water that surrounded them. To demonstrate this, a series of physical and biological measurements and sampling may be undertaken from the yacht.

In taking any sample, it is necessary to record the date and time together with the state of the tide, e.g. low water or quarter ebb. It is an advantage to compare readings taken regularly, whether hourly or even seasonally. The depth of the sample is important, as life is different at differing depths. There are five standard physical readings that should be taken: temperature, salinity, dissolved oxygen, pH and light absorption, those factors described in the earlier section on the marine environment.

Water Temperature. A thermometer with an appropriate range and a large enough scale for relatively accurate readings, preferably in a brass or weighted plastic protective case, with a loop, so that it may be safely kept or launched over the side, is ideal for taking surface readings. A bath or bathing thermometer may be used if there is nothing else to hand.

Remote reading and recording thermometers make the process of taking surface temperatures even easier. The sensor can be mounted through the hull in the way of an echo-sounder's transducer, though this means an extra hole and possible weakness. A better way, suitable for yachts with outside rudders, is to mount the sensor on the rudder post half a metre below the water line.

The depth at which the temperature is taken, of course, is important. Surface and bottom temperatures are often many degrees apart. The ordinary thermometer reacts quickly to change and it is, therefore, difficult to use an everyday instrument, except near the surface, and a more sophisticated thermometer is needed to read the temperature at a specific depth. Mercury oil thermometers used in aircraft may sometimes be obtained from surplus shops and they can be used for sub-surface readings. A more convenient answer has been provided by a number of American companies, such as Ray Jefferson of Philadelphia, for the sport fisherman, who has become temperature conscious. The angler employs the thermometer to discover the temperature layer for his particular species. These instruments are very easy to use and inexpensive.

The Negretti and Zambra or the Richter and Nansen tipping thermometers are even more sophisticated. The tipping thermometer such as that made by Kurt Gohla works on the same principle as the Nansen tipping bottle which is similarly employed for collecting water samples at predetermined depths. A messenger goes down the wire releasing the bottle or thermometer which turns nearly upside down, closing the bottle and so trapping the sample. A deep sea tipping thermometer works in a similar way but when upset, the mercury thread is broken at the constriction point. The thermometer then records the temperature at that depth (see fig. 23).

Fig. 23. Tipping thermometer and frame.

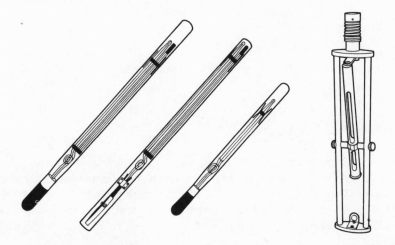

For shallow water, it is possible to show extremes of variation by using an ordinary garden maximum/minimum thermometer, suitably weighted. This should be read just below the surface or the air temperature will alter the value. However, for ordinary work there is a homemade solution that will both take water samples and enable fairly accurate temperature readings to be taken at the same time.

To make a water sampling bottle, a thermos flask, suitably protected and weighted, is fitted with a cork and ring. Protection of the glass may be done in the best traditions of the marlin spike sailor, covering the outside of a replacement thermos liner with French hitching using line made from artificial fibre. The traditional tarred hemp would be unsuitable as it may contaminate the water sample. The apparatus is lowered to the desired depth shown on a marked sounding line which is graduated to begin at the mouth of the sample bottle. The cork line is tugged gently so removing the bung, which should only be lightly pushed in or removal will be quite difficult at depth. The cork pulled out, the bottle is allowed to remain at that depth for ten minutes and then raised carefully. The contents can then be measured for temperature and salinity as well as other qualities described later. To seal the bottle completely, once full, a rubber ball and a self-closing elastic arrangement attached to the flask's harness can be used (see Fig. 24).

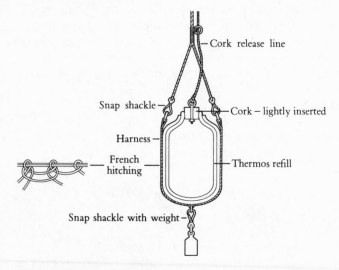

Fig. 24. Water sampling bottle.

Salinity. Salinity can be measured by chemical and physical methods. Silver nitrate titration is a very accurate way but not suitable for a small yacht. Reading the hydrometer in the cockpit of a small boat can present problems too, but if the sample is taken in a deep and narrow transparent container, it is possible to take readings and plot changes. Salinity, temperature and density are closely related. If

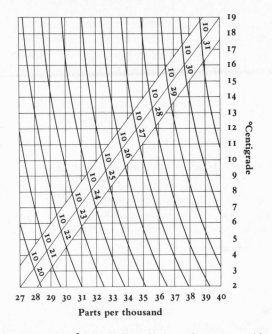

°Centigrade

Parts per thousand

Fig. 25. Admiralty chart for finding the salinity of seawater, if you are equipped with a hydrometer and thermometer. Example: temperature—11°C; hydrometer—1026; salinity—33.7 ppt.

two are known, the other can be calculated. The British Admiralty publish a chart (Misc. 354F) which may be used to discover salinity, with the help of a thermometer and hydrometer (see Fig. 25).

Much the easiest way, of course, is to use a meter which measures electrical conductivity. Such instruments often double as a thermometer and both readings can be taken with the flick of a switch.

Dissolved Oxygen. The measurement of gaseous oxygen in the water surrounding small boats is best and not very expensively done with an oxygen meter. The Winkler laboratory method is unsuitable for use in small boats.

pH Measurement. Determining the hydrogen ion concentration can be done very conveniently with a pH meter but there are water-testing tablets which can be used in seawater, though they are not as accurate, the reading being taken by comparison with a colour card.

Light Absorption. Light is important to marine life and the depth of penetration shows how far down that light is available. This is most simply measured with a Secchi disc, a white painted circular metal plate, a dustbin or pail lid suspended on a marked line. It is lowered over the shaded side of the yacht, until it disappears and a note is made of the depth. A little more line is paid out and then the disc is recovered slowly, noting the depth at which the disc reappears. The mean distance between the two is then recorded.

A photometer or an environmental comparator (Griffin & George) are more sophisticated instruments for measuring light absorption. The latter also contains, in addition to the light meter, a colorimeter and thermometer and so it can be used for comparing environments on both land and in water.

So much for methods of assessing and plotting the physical environment. It is now time to look at ways of studying life in the sea.

Biological Sampling at Sea

The equipment needed to capture a plankton sample is both simple and easy to use. Chapter Three looked at the fascinating world of plankton in some detail. Plankton studies on a small boat can be a consuming pleasure, even if no conclusions are to be drawn. It is a way of removing the blind from another natural window.

Plankton Nets
Plankton nets were formerly made of silk, but nylon is so much more durable and easier to stow that it is now universally employed. The mesh size is critical, for it determines the size of animal retained. For the yachtsman, a medium mesh or net with 300 meshes per inch will be the best bet for stowage space will be at a particular premium. The Marine Biological Association of the United Kingdom at Plymouth, for example, supplies six standard nets with ring sizes of 12" and 18" diameter, as below.

	m.p.i.	*aperture* (μ)
Coarse	26	1050
Medium	49	348
Fine	96	170
Very fine	155	90
Extra fine	192	61
Ultra fine	450	20

Plankton hauls can be carried out horizontally or vertically. The horizontal haul captures the plankton of a particular layer of water by being towed at a set depth, while the vertical haul filters the 'water column'. The speed through the water, too, is important for it maintains the filtering action. The best speed is under three knots for a horizontal tow and one metre per second is about right for the vertical haul.

The net is rigged in two different ways for the two types of haul. The horizontal employs the net as shown in Fig. 26. The vertical haul needs to be loaded differently and the weight is suspended from the net ring by three lines clear of the bucket.

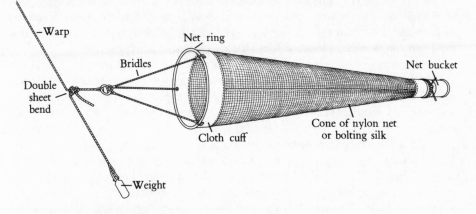

Fig. 26. A standard plankton net.

The horizontal tow can be a simple one behind the yacht, ensuring that the bitter end of the warp is properly secured. However, it is much more convenient to use a special winch.

The best is a horizontal drum, rather like a large edition of those used in Norway and Iceland for trolling. An ideal answer is a modified American 'downrigger' system favoured by the sports fisherman, such as the one made by 'Big Jon' (see Fig. 27).

Fig. 27. 'Big Jon' downrigger.

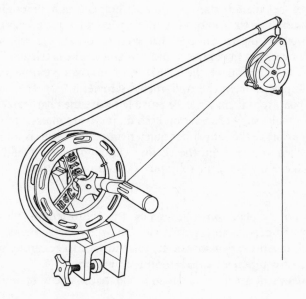

The weight, usually up to ten pounds, takes the net down to an exact depth and the winch makes recovery easy. A modification is needed in securing the net to the weight line, for in fishing the sharp tug of a fish disengages the angler's rod line from the weight, allowing him to play the fish unencumbered.

The length of time taken on the tow depends on circumstances. A bay or estuary may dictate that the tow can only be carried out on one tack. In the open sea, or under power, such limitations may not apply and a half-hour run be practicable. The important point is to sample a particular area of water and not to include estuary and open sea together, as it will not be possible to discover which species came from where, without a good deal of experience.

The plankton are filtered to the aft end of the net and concentrated in the bucket. This can be removed by loosening the cod end 'lashing' or by folding back the net and pouring the contents into another container.

Schlieper in his *Research Methods in Marine Biology* says, when talking of the 'examination of the material': 'The examination of living plankton is usually only possible immediately after the sample has been obtained. The complete fullness of form which is revealed to the observer under the binocular microscope or simple microscope and which also has great aesthetic attraction, is unfortunately of a very temporary nature. The delicate organisms, some of which have been already damaged during collection, can only tolerate the rapid changes in environmental condition (such as too high a temperature, excessive light intensity under the microscope and oxygen deficiency) for a short time and then die.'

A thermos flask may help keep the temperature down but it is best to examine the catch before the colours fade. Microphotography aboard a small vessel presents problems but it is well worth sketching interesting plants and animals. The latter may be anaesthetized with menthol chloralhydrate or propylene phenoxetol. Division of the catch, keeping part for subsequent examination, is a good idea. A stempel pipette ensures that such subdivision is representative; it consists of a plunger that fits into a round container. The catch should be shaken to ensure even distribution of the plankton. A known volume may then be removed by the pipette from the well-mixed sample.

On a small boat, too, it may not be possible to use the chart table, for the sea movement may make such observation both difficult and messy, in which case it is important to preserve the sample for quieter times. Formalin is usually a 40 per cent solution when bought from the chemist. Enough water should be added to the catch to achieve a 4 per cent solution, a ten-times dilution.

Microscopy

Looking at most of the plankton requires a microscope or lens of at least 20 × and so the ship's microscope is an essential aid. Binocular instruments are ideal for land but difficult to stow safely aboard. It is also unwise to go sailing with such an expensive piece of equipment, so selection is important.

The ideal instrument for the yachtsman is one that is able to scan the catch and then, using a different power, examine it in some detail. A hand lens, in addition, is best for the first look. For this purpose, the lens should be at least 10× and the ideal is a combined 10× and 20×. The lens should be held close to the eye and the object or specimen tube containing the catch be brought towards the observer until the specimen is in focus. Both eyes should be kept open when using a lens, or for that matter a microscope or telescope. Concentration on the object, in each case, makes viewing easy with a little practice.

To return to the microscope, low power is used for scanning and looking at larger objects. The zoea of the porcelain crab, *Porcellana platycheles* Pennant, for example, is nearly 2 mm long. Higher powers are useful for identification. The detailed features of zooplankton must be carefully examined before final identification can be relied upon.

For a yacht there are other considerations, for, as has been said above, equipment must be easily stowable and inexpensive, as the damp atmosphere of a small boat will eat into most things. To prevent damage when not in use, the microscope must have a strong and preferably airtight, but at least watertight, case. Darwin's dissecting microscope, when not in use, was protected by its stand which was in the form of a wooden box. A modern dissecting microscope is still useful for viewing the catch, but the best general purpose instrument I have come across is made in Czechoslovakia. It is sold under the name of Meopta, Praha. The microscope is contained within a waterproof metal-domed case with carrying handle. The base of the box is the foot of the microscope. The draw tube unscrews from the body and can be stowed in a socket on the foot, so reducing the height of the case to 16.5 cm. with a width of 13 cm. The instrument in its case weighs under 2 kg and can, therefore, be carried inexpensively and conveniently by air, if joining, or returning from, a yacht overseas.

This microscope has a triple nose piece, containing three objectives, which with the 15× wide-angled eye piece gives the following powers:

$$15 \times 3 \cdot 3 = 49 \cdot 5$$
$$15 \times 6 \cdot 7 = 100 \cdot 5$$
$$15 \times 20 = 300$$

Such advantages enable the sailor to observe virtually all that he would wish to see.

A 'camera lucida' fixed to the eye piece makes sketching a simple matter. The viewer, looking into the microscope, spots the outline on the piece of paper placed alongside the foot, allowing an outline of the specimen to be traced, the details being added later. Perhaps easier to use, though, is a micrometer ruled eye piece and squared paper. Neither of these useful attachments, of course, comes with the microscope; they have to be obtained separately. Observation and sorting is best achieved by removing a sample with the stempel pipette to a large petri or small pyrex dish, pressed from the galley.

First, the larger animals, such as small shrimps, worms, jellyfish, cladocera and ostracods are removed with tweezers to a watch glass. The smaller plankton may be examined by taking a sample with an ordinary pipette or an old-fashioned pen filler and placing them in a small petri dish or pond life cell. These two alternatives are better than a slide on a small boat, as a slight movement will cause the water to wander, carrying with it the object of interest. One or two animals of plankton are chosen from the dish for further study, again with a pipette.

Temporary Mounting

Specimens may be mounted temporarily by placing them on clean cavity or ring slides. If very small, a plain slide will do and Fig. 28 shows how the specimen may be correctly positioned under the cover slip, using the paper diagram of the slide as a template. The coverslip holds the specimen in place after it has been contained in a drop of water. Too much can be removed with a paper towel, too little can be made good by placing a drop alongside as shown in Fig. 28.

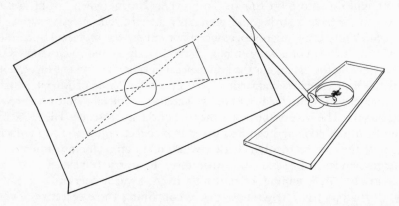

Fig. 28.   1. Centring a slide. Draw round a slide on a piece of white card. Join the corners together and so find the centre of the slide. Centre a cover slip and draw around that. Use the card for centring specimens. 2. More liquid may be introduced with a glass rod, placing the drop right alongside the cover slip.

The method of applying the cover slip is important, or bubbles will be trapped underneath the glass. The slip is tilted on, placing one edge on one side of the drop, supporting the other edge of the cover slip with a probe or mounting needle, then lowering it slowly to spread the liquid beneath the slip. If air is accidentally trapped, it may be removed by heating the needle and placing the point on the cover slip directly over the bubble.

A very convenient piece of equipment for quick viewing is the Rousellet Compressorium. The drop of liquid containing the specimen is gently squeezed between two glass plates in a brass corset, by means of a simple screw.

Permanent Mounting

The advantage of fluid mounts is that they encapsulate the specimen in what appears to be its natural environment. With the correct preserving fluid, the colour and form of fragile exhibits may be retained. Kollofast solution A for fixing followed by B for permanent mounting is excellent. This can be used in place of formalin.

There are many ways of mounting selected examples of the catch permanently, but the way given below is selected as the easiest and least demanding for the yachtsman/naturalist and his chart table/laboratory bench. (More complicated, perhaps more permanent, and certainly more time-consuming methods, may be found in the books on microscopy in the Bibliography.)

The specimen is encapsulated by a cover slip in much the same way as the temporary mount, though this time the animal or plant is always preserved in a fixing solution such as formalin or Kollofast. With zoo- and bulkier forms of phytoplankton, the appearance will be distorted if there is insufficient room under the cover slip for the specimen. Cavity and cell ringed slides overcome this problem, the latter giving the most room as the thickness of the ring can be increased from shallow to deep by cementing on more rings. For the yachtsman, it is more convenient to carry a box of cavity and already prepared cell ringed slides and cover slips to match, rather than making up the cells aboard.

Specimens are usually mounted in dilute formalin solution in the manner described for temporary work. Patent mounting fluids have advantages, particularly in preserving the colour of the specimen. Their selection is best discussed with the local naturalist supply shop. Formalin, however, is very suitable on board a yacht, although its smell is not particularly pleasant. It should not be used for minute shells as the formic acid, unless neutralized, will dissolve much of the specimen. This may be done by saturating the diluted formalin with calcium carbonate, or better still, hexamine.

The cavity or cell slide now has to be prepared so that the cover slip may be sealed in a watertight way on top of the ring or cavity to prevent evaporation of the solution containing the specimen. A ringing table makes this an easy job. This is a circular platform mounted by ball race on a heavy base, so that it may be spun. The slide is centred on the table's 'rings', which allow the slip to be properly positioned in the middle of the slide and held by clips. An artist's brush dipped in some ringing cement allows neat application to the top of the cell.

The preserving solution is then dropped into the cavity and the animal or plant transferred to it by tweezers. In the case of small animals, they are contained within the liquid and dropped by pipette. The cover slip is then applied to the cemented ring, using the method already described that avoids trapping bubbles of air. The slip is pressed firmly into the cement, removing excess liquid with a paper towel.

A protective ring that overlaps both the cover slip and the edges of the cell ring may then be cemented to the slide. The final touch is given by a decorative ring which is painted on in coloured, quick-drying shellac, again employing the ringing table.

All that remains is to label and store the slides. The label may be laid out as below on a standard slide label, available from naturalist suppliers.

Marine Plankton
  Zoo Forms
Nab Tower, Spithead
  Date:
Yacht Gang Warily R.C.C.
  Ref:

The reference could refer to the log page or to a special notebook which records physical readings such as information on tide, weather, sea state and sea and air temperatures. The label is stuck to the left-hand side of the slide, looking down.

Fluid mounts should be kept dry and stored out of the sun and heat. A slide box is essential if they are to be preserved as a record, or comparison, for any length of time, aboard or ashore. Slide boxes are available in wood to hold 50 to 100 and in card with wooden separating racks for smaller numbers.

Shore Collecting

The yachtsman bent on discovering the shore should always go well prepared. Warm clothing, boots or strong canvas shoes are usually essential. The other equipment should include:

A pocket compass. A good pocket compass is essential when wandering the shore. Darkness or fog may obscure the beach and make return difficult. The problem may become worse when the tide turns and the sea covers the sand, hiding even that reference.

A strong sailor's knife.

Spade. One of those that used to be seen attached to the side of a jeep has always seemed useful to me, provided it is brightly painted (to protect against loss) and lightly oiled.

A collecting net. Strong but small with a tough wire 'D' frame (see Fig. 29).

Plastic bags. These should be of the self-sealing and reuseable type. They can be used as temporary aquaria.

Collecting jars. The plastic milk bottle holder which has space for four or six bottles is ideal as a carrier. Suitable wide-mouthed containers of 8 cm width and under fit well and act as useful collecting jars.

Hand lens. This should be of 8× or more. The combined 10× and 20×, recommended for initial plankton surveys, is ideal. The lens should be attached to a lanyard as it is easily lost. A most convenient fitting is the swivel and yankee clip used by fishermen for easy bait changes.

A vasculum. The light-weight aluminium anodised vasculum is invaluable. The one I have in mind has a flat base for carrying jars with a lid that can be opened and closed with one hand. The dimensions are 38 × 8 × 13 cm. It has an

Fig. 29. A 'D'-shaped collecting net (wooden handle to be inserted).

adjustable shoulder strap for easy carrying. The collecting jars suggested for the milk bottle carrier should fit the vasculum and, therefore, the 'milk carrier' can remain as a reserve in the dinghy. An alternative, if only animals are to be collected, is the 'Insulex' flat jar, a wide-mouthed vacuum-insulated container which can be equipped with a battery powered aerator described later. The bait container, also described later, is useful for collecting too and may be similarly equipped with an aerator.

Specimen tubes. 75 mm × 25 mm wooden pocket-size holders are available or can be made.

A folding quadrat. This is useful for population counts of a defined area.

A sounding line. With lead and marked in metres.

A pocket thermometer. For taking the temperature of air, ground and water.

pH papers. To measure both the salinity and alkalinity of the water and the substrata.

With this equipment, it is possible to learn a good deal about the shore. It is best to walk first to the top and note what is to be found on the strand line. Litter, too, gives a clue. A mass of plastic cups, for example, may indicate a poorly run ferry service where the crew are allowed or, indeed, encouraged to dispose of customers' cups by the sackful straight over the stern. One or two could perhaps come ashore from a yacht or passing ship and are normal flotsam and jetsam.

Shells around the strand line indicate, in a rough fashion, the animals living in the zones below or drawn from those that lie to windward. It is now time to turn paper knowledge into experience by using the notes in Chapter One and looking down the beach exposed by the low tide. Taking a rocky shore, note how the beach is zoned and search out the green seaweed zone, species of *Enteromorpha* and sea-lettuce, *Ulva lactuca*, the brown seaweeds, the wracks such as bladder, *Fucus vesiculosus*, and the oarweed, *Laminaria digitata*, on the extreme lower shore.

It is important to turn over the stones to see the animals sheltering beneath and then recognise how they are banded in the same manner as plants.

The Ship's Aquarium

Collect sparingly for the ship's aquarium, labelling the plastic bags, containers or specimen tubes with the zone and trying to identify the specimens so gathered. The animals that live reasonably happily for short periods in an aquarium are crabs, prawns, shrimps, winkles, whelks, barnacles and sea-anemones. The latter have a particular and perhaps peculiar history of success, as can be seen from this 19th-century newspaper cutting.

A Curious Pet

Speaking on 'The Fish of the Sea', at the Royal Institution, Professor D'Arcy W. Thompson told how Sir John Graham Dalyell had kept a sea-anemone as a pet. The old Scottish naturalist took one of these sea-flowers home and fed and cared for it so well that it grew and produced little ones. He named it 'Granny', and when he died it went into the keeping of another naturalist, who also passed away; and 'Granny' successively came into the possession of an Arctic explorer, a schoolmistress, and a botanist. The sad part of the story was that 'Granny' was so neglected by her last owner that she died also, after spending sixty years in her dish of water, and gaining such renown that she was accorded a half-column obituary in *The Scotsman*.

Both the collecting containers and the ship's aquarium require oxygen. For the yachtsman/naturalist though the oxygen problem is short-term, spanning at the most a few days while at anchor. The green weeds can help low populations but more sophisticated arrangements are to be preferred. The Seaway Corp. of Taiwan make for the American market an ideal arrangement for the yachtsman/naturalist. It is a small 1·5 battery-powered aerator, working on an SP 2 (D battery) torch cell. The plastic case, which can be used to stow the air line and weighted diffuser, measures 13 × 7·5 × 4·0 cm and weighs 252 g complete with battery. This, with a 0·57-litre 'Insulex' snack jar, will keep specimens at sea temperature on warm days, if a small hole for the air tube is drilled in the top. The Americans also make an 'Old Pal' minnow bucket which is useful for this purpose. They are constructed of polyethylene, float upright, are unsinkable and self-aerating (see Fig. 30). These buckets can be slung over the side of the dinghy on their retaining line or anchored with a stone or some other weight in a few inches of water.

Once aboard, aeration can be continued with the battery-powered pump but it is noisy and expensive on cells. The yacht should have its own aerator working off the ship's batteries. Marine aquarium pumps are of low consumption but usually use the main supply. This difficulty can be overcome by fitting a transverter. One such model is made by Valradio of Feltham, near London. The Americans have produced a useful arrangement that is able to aerate two live wells simultaneously. The amount of air entering the tanks can be adjusted by a flow control. The equipment is made by Jabsco Products ITT of California.

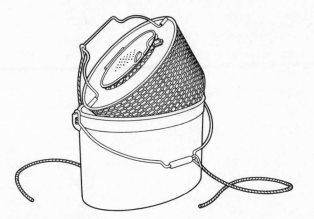

Fig. 30. A minnow bucket.

The rock pool is a particularly interesting place to search for animals for the aquarium, sweeping upward amongst the weeds with the net to search for creatures that live amongst the fronds (see the notes in Chapter One). The tufts of corallines (*Corallina officinalis*) are often inhabited by small starfish. Fish like the spined stickleback (*Spinachia spinachia*) are sometimes left in the pool by the tide. Remember the holdfast or root of the oarweed is the home of a vast number and variety of animals and is well worth close examination with the hand lens.

Bivalves such as the cockle (*Cerastoderma edule*) of the lower shore, shut tight when handled, but will show their siphons if given the peace of an aquarium. Their burrowing action into the sand can be watched, if some are brought aboard in a container. The sand should be spread to at least the depth of the 'length' of the shell on the bottom of the aquarium, so that this submerging action can be watched.

A way of ensuring that dead bivalves are not carried aboard, either for the aquarium or for the pot, is to hold them between the thumb and forefinger and push the right valve one way and the left the other. If the animal resists, it is alive; if the shell divides to reveal a muddy interior, the specimen should be discarded and the search for others should be continued. As the animals under study are only to be in the ship's aquarium for hours or a couple of days at the most, it is not worth considering feeding them. It is important though to watch for any that are dead, and remove them, or the water will become foul and promote further losses.

The general health of animals and plants in the aquaria is promoted by keeping the container out of the sun while not observing or photographing. The aquarium should, if possible, have its own stowage space in which it can sit while charged. A baffled top should prevent spillage when anchored in a seaway, or making a short passage, though this should be rarely contemplated while the aquarium is full. It is possible to combine this stowage with a bulkhead box that can be

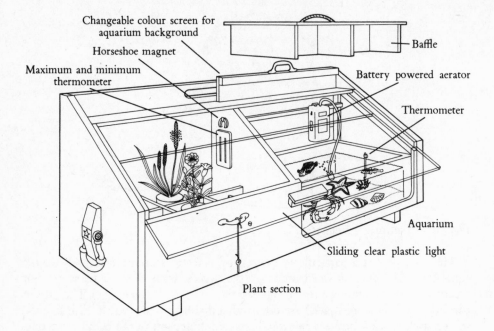

Changeable colour screen for
aquarium background

Horseshoe magnet

Maximum and minimum
thermometer

Baffle

Battery powered aerator

Thermometer

Aquarium

Sliding clear plastic light

Plant section

Fig. 31.   Plant case and aquarium with baffle removed.

removed and taken to other parts of the ship for study, or ashore. The box has a
section devoted to keeping plants (see Fig. 31).

It is infinitely preferable to put the plants and animals back where they were
found a short time after the study, or when photography is completed.

Shore Recording

Chapter One described the coast in detail under the headings of rocky, sandy and
muddy shores. With this information as a background, it is possible to learn a
great deal more about the plants and animals that live in these different habitats
and to note the variations within them. To draw any conclusions, it will be
necessary to measure and note.

The lead line, marked in fathoms or preferably metres, is ideal for measuring
the way down the shore from the splash zone to the low tide level; the lead itself
acts as an anchor for one end of the line. The actual 'line' down the shore is
known as a 'transect', or to be more accurate a 'line transect'. Using this as a
guide, all algae touching the line may be then recorded, all animals within half a
metre of the rope put in the notebook too, so that eventually the log is a record of
this shore and can be then compared with similar types.

A 'belt transect' uses the line transect and employs a 'quadrat' for counting the number of animals and plants within each square all the way down the beach, square upon square, or say at intervals of ten metres, depending on the type and scale of the beach, but also perhaps the number of crew employed in the work. One member may be responsible for recording the algae and another, the animals. The result should end up as names recorded down a line, as in Fig. 32.

A check list is a quicker way of achieving the same results and blanks can be roneoed or photocopied in advance for use on different shores and for either animals or plants.

Temperature and salinity should also be measured with the thermometer and the test papers, noting the readings and particularly the variations that will be seen if measurements are taken down the shore.

Rock pools may be mapped, using the quadrat. The plant and animal populations are recorded and charted according to each square, a grid of the pond being made, each square of quadrat size, with spare line and pegs, before the study begins. This grid will also help in mapping the pond.

Fig. 32. A line transect.

Date	Yards along transect	
Temperature	**Animals**	**Seaweeds**
		■ **High tide mark**
	Littorina 1	*Ulva* sp
	saxatilis 2	*Enteromorpha* sp
	3	*Ulva* sp
	4	*Cladophora* sp
	5	
	6	
	7	
	8	
	9	*Fucus spiralis (Spiral wrack)*
	10	*Fucus spiralis*
	11	*Fucus spiralis*
	12	
	Patella sp 13	
	(Limpet) 14	
	15	
	Littorina 16	*F. vesiculosus (Bladder wrack)*
	Littoralis 17	*F. vesiculosus*
	18	*F. vesiculosus*
	Nucella 19	
	(Whelk) 20	
	21	
	22	*F. serratus (Saw wrack)*
	23	
	Balanus sp 24	
	(Barnacle) 25	
	26	*Chondrus* sp
	27	
	28	*Laurencia* sp
	Carcinus sp 29	
	(Crab) 30	
	31	*Laminaria* sp *(Oar weed)*
	Asterias sp 32	
	(Starfish) 33	
		■ **Low tide mark**

The species must be clear in this illustrated map and a key developed; the actual symbols are the choice of the artist. When this is transferred to the journal, or log, it may be done in colour, to enliven the record.

Plant Collecting

In almost all circumstances plants should be left where they are, to be photographed and enjoyed. The British Nature Conservancy Council exhorts 'bring the book to the plants, not the plants to the book'. It is easier, in any event, to identify a plant and animal in its typical surroundings, than when viewed naked on the saloon table. However, very occasionally, it may be necessary to remove plants as habitat for the animals that depend upon them, for the aquarium or for particular study. These should be common plants, in abundance, and not taken, in any circumstances, from nature reserves.

The Wardian case, invented by Dr Nathaniel Bagshaw Ward in 1842, was originally designed to show that animals and plants might be kept in airtight glass cases and that each might be so adjusted to 'breathe' in what the other 'breathed' out. The Wardian case does not need hermetic sealing. It is, in effect, a little greenhouse: it lets in the sun; water vapour is unable to escape and so condenses on the contents and the panes, dripping down in the cool of the evening, on the soil or moss below. This cycle continues daily and so no further attention is required for a period extending for weeks or even for months.

The Wardian case has a place on a small yacht. With a plastic or reinforced glass top, it may be placed on the coach roof or perhaps aft, behind the cockpit coaming. But such elaboration is to approach perfection and the plastic bag may be considered the modern Wardian case. These may be used inside the aquarium and plant case, or the bags may be placed in racks, in the light, either inside the coach roof, or around the skylight in older yachts.

Again, each specimen must be labelled with the date and location (latitude and longitude) of collection and the name, if known. If unsure of the name, decorate with a question mark. Neighbouring plants may be noted as well, for they help in the final diagnosis.

Seed Collecting

Most people collect seeds before the fruit is ripe. In Europe most seeds are ready between October and the end of December. The vasculum is useful for collecting both plants and seeds, though it is important not to bring aboard too much leaf and stalk, as when dry the harvest has to be winnowed out. This is more conveniently done ashore using a flat white sheet of paper, anchored to the ground, with a light breeze to shift the chaff. A shady place on a sunny day is the best spot for drying the pods and seeds. The artificial warmth of the oven is to be discouraged.

Preserving Specimens

The vasculum is useful for collecting plants and their leaves and fruit, but an added aid is the 'collecting portfolio'. This is a large book or folder that keeps plants in better condition than the sometimes crowded confines of the vasculum. It is large, but could find a home among the spare charts on a larger yacht.

The 'collecting portfolio' is constructed of two sizeable sheets of cardboard, usually 60 × 30 cm, bound along one side with canvas in the way of a spine. Drying paper is placed loose within the covers, as pages in a book, and the whole is carried on the shoulder as a bag by a webbing strap, one secured to the top and the other to the opposite bottom corner of the boards. The cover is thus kept closed by its own weight. In due course, the plants from the collecting portfolio are transferred to the plant press.

The plant press is a way of permanently preserving specimens, and small presses, ideal for yachts, are now available. A typical set consists of two small plywood boards, 18·5 × 18·5 cm, tensioned together at each corner with a butterfly screw. The sandwich filling between the two consists of sheets of corrugated cardboard, interleaved with two sheets of tawny coloured absorbent drying paper. The leaves, flowers and (if not too fleshy) the fruit are arranged between the absorbent sheets and they, in turn, are placed between the cardboard spacers. The press is then evenly tightened with the four screws and the specimens left *in situ,* for as long as possible. With fleshy plants, the paper should be changed daily until sufficiently dry to be left. At least a month is preferable before mounting on special paper, securing the specimen with herbarium paste or sticky tape.

Fruits may occasionally show signs of mildew in the warm, damp atmosphere of the boat, particularly when there has been a spell of wet weather. This problem may be overcome by brushing lightly with a solution made from tablets of 'corrosive sublimate' (mercuric chloride), dissolved in methylated spirits. This compound of mercury is dangerous and should be handled with care.

Larger presses, 43 × 28 cm, made of plastic-covered metal frames are available for the more serious collector. Pressure is applied to this press by two web straps. It is possible to make your own press of similar dimensions with wood lath in the form of a grating, using book straps to apply pressure.

Plant presses preserve the colour and form of the plant if properly used and the result is both useful for identification and decorative.

Dry, pressed material can also be inserted between two self-adhesive sheets of 'Fablon', a clear plastic film used for lining household drawers. Fresh specimens, treated in this way, will almost certainly turn black in a short time. Encasing in plastic enjoys an advantage, for it allows both sides of the leaf to be inspected and the specimen remains flexible, can be shown to children and subjected to rough treatment without damage.

Seaweeds may be pressed, too, though the technique is necessarily different. The plant is placed in a tray of seawater and mounting paper is put into the water under the specimen. The fronds are arranged with a paintbrush to the best advantage and then the paper and the plant are lifted out carefully together and blotted to remove excess seawater. The specimen is put into the press and the paper changed daily, as for fleshy plants, until the seaweed is dry.

Seaweeds contain their own mucillage which acts as glue, but it may be necessary to reinforce this with the normal herbarium paste or strips of gummed tape for final display or storage. Old-fashioned sellotape may turn brown and brittle after a time and, therefore, should be avoided.

Shell Collecting

Collecting shells has fascinated sailors for centuries, and is usually a favourite pursuit of children on their first visit to the beach. The strand line is often littered with the discarded shells of dead molluscs. The most dedicated of collectors will not pay much attention to this area, preferring to collect specimens complete and, therefore, alive (bearing in mind, of course, that in the interests of conservation, his haul should be modest). Discarded shells on the beach give little idea of when the late animal lived and, therefore, the reasons for the type of shell and its coloration. In many cases coloration is lost by weathering.

To learn about the animals that built the shell, it is obviously important to see them alive and in their own habitat, and Chapter One should be consulted before the shore is reached.

Collectors require only the simplest equipment. The list below is reasonably complete. It is, however, doubtful whether the collector would encumber himself with half of it, though each has its particular purpose and some of the aids are in common with those recommended for the naturalist.

Clothing. Boots, or canvas shoes. A jacket with a 'poacher's' and other large pockets. A canvas belt with military pouches would be useful. This could be made at home in the form of a money belt, with a number of folded pockets.

Carrier and containers. The plastic milk bottle carrier described before could be invaluable if armed with wide-mouthed plastic collecting jars. Division of the catch is useful, as separation helps identify what came from where, when sophisticated notes and labelling are difficult. Plastic tubes for small specimens in wet and rough conditions, should also be included in the jacket, or the belt pockets.

Labelling. A supply of pulpless labels should be carried with a pencil or biro, to ensure proper records later. Pulpless labels do not disintegrate when wet. The collector, date, time, place, the state of tide and the habitat should be given as below:

YACHT . . . .
Collector . . . .
Date. . . . .
Time . . . .
State of Tide . . . .
AT. . . . .
Specimen . . . .
(Using generic & trivial name, see Introduction. Include whether abundant, common or rare on that particular beach.)
Remarks. . . . .

Roneoed or printed cards with this information are a great help.

A pocket compass. As recommended earlier.

A sieve or screen. A wooden-sided screen with 0·5 cm mesh is useful for sieving out small molluscs or their shells. A square-shaped one with 50 cm sides, 10 cm high, will serve well. The mesh may be of wire or nylon net, the latter better suited to storage aboard. A plastic gravy strainer is also useful for small specimens.

The look box. Such a box was discussed earlier when describing photographic methods and the same design will serve well for examining the bottom from a dinghy or when wading about. It is possible to make such a 'look box' by letting a round sheet of plate glass into the bottom of a plastic bucket and sealing with caulking compound. The glass should be a tight fit against the sides of the bucket and a retaining rim left in the bottom, forming the caulking lip. A line to the handle will stop the viewing box floating away when not in use.

Long-handled net. A long-handled net on a stout wire loop may be employed with the look box for harvesting from the rocks or the bottom.

Hand lens. 10×, 20× as recommended earlier.

Spade, fork or trowel. The latter is useful for bivalves that live just below the surface. Deeper burrowers will need the spade or fork. The entrenching spade, recommended before, is ideal for both levels.

Formalin. Small live specimens may be preserved at once, by dropping them into a specimen tube containing 4 per cent formalin, to which calcium carbonate has been added to prevent acid damage to the shell.

Dredging

Many species of shellfish live in deep water and may be collected by diving or by employing the naturalist's dredge (see Fig. 33). This is a heavy rectangular iron frame with a net fixed to the aft end. Four short chains connect to a ring and the towing warp is secured to this. The warp should be at least three times the

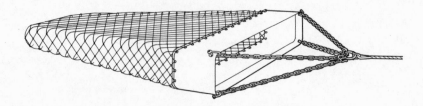

Fig. 33. The naturalist's dredge.

working depth. When operating on a rough or foul bottom, the dredge may stick and a buoyed line made fast to the cod end will help recovery. The towing yacht should proceed at slow speeds, 2–3 knots at the most, or the dredge will bump along the bottom and not fish properly. One of the most interesting ways of obtaining a catch from deep water is from the stomachs of demersal or bottom-feeding fish. The shells are often undamaged, the animal having been digested out by its captor. The interior of a fish's stomach is valuable in assessing their diet. For example, whole crabs have been found in the stomachs of thornback rays.

Once aboard, there are different ways of cleaning live specimens. One method is to wash off any mud or sand and store the shells in a plastic container filled with alcohol. This eventually loosens the animal from its shell and is suitable for both bivalves and gastropods. The specimens may be left indefinitely and final cleaning may wait until the collection is brought ashore. There are quicker methods of cleaning. Bivalves may be encouraged to open by placing them in warm fresh water. The animal will gape and the adductor muscles may then be severed with a sharp knife. Freezing produces the same result. Gastropods are more difficult and should be placed in water and brought to the boil. The inhabitant may then be extracted with a pin or sail needle. The operculum, or 'little lid', that seals the opening should be saved and replaced, secured with glue, before display.

Fishermen often kill and in so doing preserve starfish, by leaving them to sunbathe to death on the wheelhouse roof. Such treatment for these and for shellfish is not suitable for a small yacht, as the smell takes the edge off the enjoyment of being afloat.

Some collectors take their shells home and bury them in the garden to clean them. Peter Dance in his excellent book *The Shell Collector's Guide* (David & Charles, 1976), discusses the merits and the ethics of hanging large gastropods, aperture downward, from a tree branch, until they drop out.

Whatever method is employed, the shell should be scrubbed and left for a few hours in a very dilute solution of household bleach. With gastropods, it may be advisable to finally flush the shell with 70 per cent alcohol or even vodka and plug for a few days with cotton wool, leaving the shell, aperture upwards, in an egg box. This will help to fix any small pieces of the animal left behind during the

Plate XI

FISH, CEPHALOPODS AND CETACEA OF THE OPEN SEA

1. *Cetorhinus maximus*: Basking shark
2. *Isurus oxyrinchus*: Mako
3. *Lamna nasus*: Porbeagle or mackerel shark
4. *Alopias vulpinus*: Thresher or fox shark
5. *Prionace glauca*: Blue shark
6. *Galeorhinus galeus*: Tope
7. *Sardina pilchardus*: Sardine or pilchard
8. *Clupea harengus*: Herring
9. *Acipenser sturio*: Sturgeon
10. *Torpedo marmorata*: Marbled electric ray
11. *Salmo solar*: Atlantic salmon (male, in breeding colours)
12. *Belone belone*: Garfish
13. *Zeus faber*: John Dory
14. *Gadus morhua*: Cod
15. *Pollachius pollachius*: Pollack
16. *Dicentrarchus labrax*: Bass
17. *Chelon labrosus*: Thick-lipped grey mullet
18. *Pollachius virens*: Coalfish, saithe or coley
19. *Cyclopterus lumpus*: Lump sucker or sea-hen
20. *Trachurus trachurus*: Horse mackerel or scad
21. *Mullus surmuletus*: Red mullet
22. *Scomber scombrus*: Mackerel
23. *Lophius piscatorius*: Anglerfish
24. *Aspitrigla cuculus*: Red gurnard

25

26

27

28

29

30

31

32

33

34

36

37

Plate XI

FISH, CEPHALOPODS AND CETACEA OF THE OPEN SEA

25. *Scophthalmus maximus*: Turbot
26. *Sepia officinalis*: Common cuttlefish
27. *Scophthalmus rhombus*: Brill
28. *Pleuronectes platessa*: Plaice
29. *Platichthys flesus*: Flounder
30. *Mola mola*: Sunfish
31. *Solea solea*: Dover sole
32. *Loligo vulgaris*: Long-finned squid
33. *Globicephala melaena*: Pilot whale or blackfish
34. *Balaenoptera acutorostrata*: Lesser rorqual or pike whale
35. *Delphinus delphis*: Common dolphin
36. *Phocaena phocaena*: Common or harbour porpoise
37. *Orcinus orca*: Killer whale or grampus

Plate XII

SEA AND COASTAL BIRDS
(All birds are in summer plumage)

1. *Larus fuscus*: Lesser black-backed gull
2. *Rissa tridactyla*: Kittiwake
3. *Larus marinus*: Greater black-backed gull
4. *Larus argentatus*: Herring gull
5. *Sterna albifrons*: Little tern
6. *Larus canus*: Common gull
7. *Fulmarus glacialis*: Fulmar
8. *Sterna dougallii*: Roseate tern
9. *Sterna hirundo*: Common tern
10. *Alca torda*: Razorbill
11. *Sterna sandvicensis*: Sandwich tern
12. *Uria aalge*: Guillemot
13. *Cepphus grylle*: Black guillemot
14. *Fratercula arctica*: Puffin
15. *Gavia stellata*: Red-throated diver
16. *Phalacrocorax carbo*: Cormorant

17

18

19

20

21

22

23

24

25

26

27

28

29

30

Plate XII

SEA AND COASTAL BIRDS
(All birds are in summer plumage)

17. *Sula bassana* : Gannet (inset picture shows the method of diving for fish with half-closed wings)
18. *Puffinus puffinus* : Manx shearwater
19. *Hydrobates pelagicus* : Storm petrel
20. *Aythya marila* : Scaup (male)
21. *Anas penelope* : Wigeon (male)
22. *Melanitta nigra* : Common scoter (male)
23. *Somateria mollissima* : Common eider duck (male)
24. *Mergus serrator* : Red-breasted merganser (male)
25. *Stercorarius parasiticus* : Arctic skua
26. *Falco tinnunculus* : Kestrel (male) hovering
27. *Stercorarius skua* : Great skua (male) in threat display
28. *Pyrrhocorax pyrrhocorax* : Chough
29. *Pandion haliaetus* : Osprey, viewed from below. The upper side is blackish-brown
30. *Sturnus vulgaris* : Starling

extraction process. Adhesions left on the outside of the shell, after the bleach, may be removed with a sharp knife. Some of these barnacles and worm tubes could be interesting and should be left.

When thoroughly dry, the colours may be enhanced by a thin coat of varnish applied with a soft brush. All that is now left is final labelling and storage. The label should follow the one already described. Cardboard boxes and cotton wool, tissue or kitchen paper are useful for packing.

Final display ashore may take a number of forms. The old-fashioned cabinet will preserve valuable collections for ever, but it is the cabinet that decorates the room, not the shells. Glass-topped boxes with divisions are attractive and common shells look well half or completely embedded in resin. An ice-cube tray makes a useful former for this. A layer of resin is allowed to set, the shells are arranged and labels inserted before the second and retaining layer is added. The result may be framed and used to decorate the saloon or the house.

A sheet of black-painted plywood, the corners rounded off and the shells arranged on the surface, secured with 'Araldite' or another epoxy glue, and properly labelled, provides another decorative and instructive display.

Bird Watching At Sea

A survey of seabirds and their ways is discussed in Chapter Three. The health of seabirds is of great importance, for they are a valuable early warning system of damage to the sea environment. Monitoring is vital if we are to conserve the vast natural resource which, with new technology, we are beginning to be able to farm, as well as to continue, with conservation, to harvest. However, side by side with this goes our capability for destruction and this grows by the month, with the exploitation of both oil and mineral wealth from the sea.

The yachtsman may play a part in this monitoring process by bird watching at sea. The British Seabird Group was set up in 1965 by Dr W. R. P. Bourne to discover where seabirds nested around the British Isles and to plot their numbers as accurately as possible. The great bird cliffs of the north and west of Britain were visited and counted, as well as every known site of lesser importance. A thousand observers took part, to produce a wealth of maps and statistics. An excellent book was published as a result—*The Seabirds of Great Britain & Ireland*—by Cramp, Bourne & Saunders. Yachtsmen could continue this work and make a valuable contribution to our knowledge of the ways of seabirds for there is a great deal more to be learnt about their distribution at sea. This is particularly true of areas where great water masses meet. These are places of mixing and upwelling, where natural blooms form a broad base for the start of the food chain. The sea areas include those off the south-west coast of Ireland, the western approaches of the Channel between Cornwall and the Brittany coast, the northern North Sea that stretches from Norway, westward to the Orkneys and Shetland Islands, and the summer-rich waters between Northern Ireland and the Outer Hebrides.

The methods of recording animals and plants on the shore are used at sea and a sea transect sailed. The course need not be special, though it would be of particular value if it were in the areas mentioned above. The birds, species and numbers are recorded along the course line at, say, ten-minute intervals. Each member of the crew could take a watch, if they were able to recognise with some certainty the birds spotted. The process is continued for six consecutive ten-minute counts in the hour, followed by a break of, say, thirty minutes, for such concentration can be both tiring and absorbing, taking time and energy from the business of sailing. In place of identifying every species, interest may be taken in a particular type of bird. A beginner may soon be able to recognise the shape and flying habits of the gannet, for example. These counts may also be undertaken when at anchor, or when visiting remote stacks, backed by photographs, as already suggested.

The sheltered waters behind the German and Dutch Friesian Islands is a place for seeing and recording a multitude of birds. They come in thousands in the spring and autumn, remaining to nest and resting from their migrations. Few people are able to be closer to them and to observe their habits than the yachtsman who shares their environment, especially if he is prepared to dry out (see Chapter Four).

The information collected will usually be welcomed by national bird societies. It would be an advantage to discover beforehand the particular data that is required by an organisation—a specified aim would add further dimension to the research.

Seabirds act as an aid to navigators and a good identification guide is an essential part of the serious mariner's library. A suitable selection is to be found in the Bibliography. A classic work, certainly out of the ordinary, is *Nature Is Your Guide* by the Tasmanian pioneer navigator, Harold Gatty. His 'Raft Book' saved the lives of hundreds who found themselves adrift during the last war.

Gatty looked at the habits of seabirds and wrote that 'on the whole it is the seabird communities of the higher latitudes (both north and south) that tend to include species that range far from their breeding grounds into those parts of the open ocean most remote from land. The majority of tropical seabirds, on the other hand, do not often range far from their nesting places on island groups and continental coastlines.'

Looking at northern species through his eyes, the following observations are useful. Fulmars may be seen at great distances from land, though increasing numbers show proximity to islands or the shore. Gulls, with the exception of kittiwakes, on the other hand, are basically shore birds and are seldom seen more than fifty miles from the beach. They go further out when the continental shelf extends beyond this and gulls will usually follow ships. The line is, therefore, blurred. An increase in the number of birds seen, though, is an excellent indicator on the coming of land. Further bird observation in northern latitudes may develop the art of natural navigation, so that it becomes a useful aid to the ordinary sailor.

Choosing Binoculars

The essential aid for bird watching is a good pair of binoculars and with the multitude of these on the market, it is difficult to discover the ideal pair for such purpose. Much of the advice given below comes direct from Arthur Frank, late of the firm, and his little booklet *Tell Me Mr Frank*.

There are a number of factors to be taken into consideration in choosing a pair of binoculars. *Magnification, brightness, field of view* and *weight* are all related. *Price* is obviously an important factor, too.

For bird watching and navigational purposes, a *magnification* of between 6 and 10 should be considered. Today 8×, which is a favoured power for bird watching, is the most common choice and for combined bird watching and sailing it would be the most suitable, particularly if the object glass, or the one furthest from the eye, is about 50 mm. 8 × 50 is interpreted as follows. The 8 is the linear magnification. This simply means that the object viewed is seen 8 times larger with the binocular than with the naked eye. Conversely, an object 8 miles away would appear to be only 1 mile away.

The second figure of 50, in the 8 × 50 binocular, is the diameter of the object glass in millimetres.

The field of view is the area seen through the glasses. This is sometimes engraved on the binocular plate as an angular measurement, e.g. 8° or, as is more common nowadays, it can be given as a measurement of feet over a fixed distance of a thousand yards. Field of view is closely related to magnification, and generally speaking, the higher the magnification, the smaller the field of view. A wide field is often an advantage, as when observing birds in flight.

Image *brightness* (light transmission) is extremely important, for there is obviously no point in having a highly magnified image which is so dull that it can hardly be seen. It can be taken as a general rule that the higher the magnification, the duller the image. However, by increasing the diameter of the object glass, an increase in image brightness is gained. There is nothing complicated about measuring the image brightness of a binocular. If the binocular is held at arm's length from the eyes, a circle of light will be seen in each eyepiece. This is referred to as the 'exit pupil'. The diameter of this exit pupil can be easily calculated by dividing the diameter of object glass by the binocular magnification, e.g. $50 \div 10 = 5$. An exit pupil of 3, 4 or 5 is adequate for viewing in normal conditions, but in cases where optimum performance is regularly required, under dull conditions, a larger exit pupil may be indicated. A 7 × 50 binocular, sometimes referred to as a night glass, will provide an exit pupil of 7 mm. The reason why a binocular with a 7 mm exit pupil is suitable for use in dull conditions is that in such conditions the pupil of the eye will expand to about 7 mm diameter and so it will just accept comfortably the exit pupil of light of 7 mm emerging from the binocular. This makes best use of available light.

Weight is also a factor, for with constant use, even a few extra grams can be tiring and weary hands are unsteady, and even the best binocular cannot compensate for this.

If you feel like paying top *price*, there is no doubt that the yachtsman/naturalist would be delighted with the superb binocular made by, say, Leitz or Zeiss, or those costing rather less, but still of magnificent quality, by Barr & Stroud or Ross. However, it is true to say that several of the relatively low-priced Japanese binoculars and certain of the Russian models have a performance comparable with more expensive instruments costing three or four times as much. They can be purchased with confidence, for not only are they good optically, they are also good mechanically. A guarantee against defects arising from faulty workmanship and material is supplied with these binoculars, but it is rarely necessary to take advantage of the guarantee.

Focusing. Some binoculars are described as centre-focusing and others as eyepiece-focusing. Centre-focusing is a speedy system and is particularly recommended in binoculars with a magnification of 8 and upwards. In the centre-focus binocular, it will be found that one eyepiece (usually the right) is adjustable so that the user can make the necessary adjustment to compensate for any difference between his two eyes. If the binocular is of the individual eyepiece-focusing type, each eyepiece is focused separately. This type of focusing is slightly slower than the centre-focusing type but in relatively low power binoculars such as 6× and 7×, which have a wide depth of field, the individual eyepiece form of focusing is particularly suitable. These relatively low power binoculars are virtually 'fixed focus', and once focused there is little need to readjust them. The question of speed is, therefore, not a matter of great consequence.

Coated lenses and care of binoculars. Practically every binocular produced today has coated (or bloomed) lenses and prisms. This coating consists of a thin layer of a suitable transparent substance, usually magnesium fluoride, which is deposited on the surface of the glass. By cutting out unwanted reflection the light transmission is increased, and light scatter inside the instrument is reduced. For maximum efficiency, the coating requires to be extremely thin, only $1/4$ wave band. Constant polishing and hard rubbing will affect this thin film and, for this reason, many manufacturers do not coat external surfaces.

Such valuable equipment requires care and attention. Only the external surfaces of the lenses should be cleaned and this is done with a clean but not new handkerchief. There is seldom need to rub vigorously, and special care should be taken if the external surfaces are bloomed. If the lenses are extremely dirty and greasy, they can be cleaned with a piece of linen which has been dampened with rectified turps. The turps should be used very sparingly and when the lenses are dry they can be polished with a clean handkerchief. If binoculars require internal cleaning or any adjustment which entails the dismantling of the instrument, it should be sent to a firm which specialises in this type of work. Even the simple act of unscrewing an object-glass is sufficient to upset the alignment of a binocular, thereby rendering it unusable.

Glossary

Abdomen: Posterior region of an animal's body.

Adductor muscles: Muscles which close the shells of bivalves.

Aerofoil: A body for which, when passing through air, the resistance to motion (drag) is many times less than the force perpendicular to motion (lift).

Angiosperm: Higher plant which bears seeds in an enclosed case.

Anoxic: Deficient in oxygen.

Apical: Botanical term for tip or summit.

Artic: Most northern biogeographical region.

Atrium: Chamber, e.g. that surrounding the pharynx in Tunicata.

Auxospore: Zygote of diatoms formed by union of two individuals at limit of decrease in size.

Baleen: Fibrous, horny material in the upper jaw of whalebone whales.

Benthic: Bottom-living; occurring on the seabed.

Bivalve: Mollusc in which the body is enclosed by two hinged saucer-like shells.

Bloom: Seasonal, dense growth of phytoplankton.

Boreal: Northerly biogeographical region.

Bract: Medusa-like zooid of siphonophore.

Bronchi: Tubes connecting the trachea with the lungs.

Buffer: Salt solution which minimises change in pH when an acid or alkali is added.

Byssus: A bundle of sticky fibres produced by the foot of some species of bivalve for attachment.

Carapace: Chitinous or bony shield covering whole or part of certain animals.

Cartilage: Elastic tissue or gristle often found in connection with animal bones, or replacing bone in the case of sharks and rays.

Cell: The basic unit of living material containing a nucleus and other organelles.

Choanocyte: Cell with funnel-shaped rim or collar round the base as in sponges.

Chlorophyll: Green photosynthetic pigment in plants.

Cilia: Minute threadlike organs which beat in unison with other cilia to move an organism or obtain food.

Chloroplast: A minute granule found in plant cells which have been exposed to light—contains the photosynthetic pigment chlorophyll.

Coelom: Body cavity.

Columellar teeth: Teeth-like projections on the central pillar in gastropod shells.

Connective tissue: Tissue with large quantities of intercellular material, usually connecting and supporting other tissue.

Continental shelf: Submerged edge of continental land mass usually taken to be delineated by 2000-m isobenth.

213

Crenulations: Notching on the margin of some bivalve shells.

Cuticle: Outer skin.

Dermersal: Living on or near the seabed.

Diatom: A unicellular form of algae with walls impregnated with silica.

Dinoflagellate: Microscopic organism possessing two flagella.

Dorsal: Back of an animal, or surface part of a flower furthest from the centre.

Ecology: That part of biology which is concerned with the relationship between organisms and their surroundings.

Ectoderm: Outer layer of cells.

Epicone: Part anterior to girdle in dinoflagellates.

Epitheca: Theca covering epicone in dinoflagellates.

Eutrofication: Over-enrichment of waters with plant nutrients, either naturally or as a result of pollution.

Exoskeleton: Skeleton which surrounds soft parts of an animal e.g. in arthropods.

Flagellum: Lash-like swimming organ of certain microscopic organisms, cells and male gametes.

Food web: Interconnected food chains.

Frond: Part of the thallus of certain seaweeds.

Frustule: The siliceous two-valved shell and protoplasm of a diatom.

Gamete: Reproductive or sexual cell; sperm or egg.

Gametophyte: Gamete forming phase in alternation in plant generations.

Gastropod: Mollusc with a ventral muscular disc adapted for creeping: a snail.

Gill: Respiratory organ of aquatic animals.

Habitat: Locality or external environment in which an animal or plant lives.

Haemoglobin: The red respiratory pigment of the blood of vertebrates and a few invertebrates.

Hinge: Dorsal region of the bivalve shell along which the valves meet and where they may be held together by interlocking teeth.

Holdfast: Extension of thallus used for attachment.

Holoplankton: Animals in which every stage of the life history is planktonic.

Humic acid: Dark-coloured decomposition product of animal or vegetable matter from soil.

Hypocone: Part posterior to girdle in dinoflagellates.

Hypotheca: Theca covering hypocone in dinoflagellates.

Isobenth: Line on a chart connecting places of equal depth.

Isostatic equilibrium: A condition of equilibrium held to exist in the earth's crust.

Larva: An embryo which becomes self-sustaining and independent before taking on the parental features.

Macrophyte: Multi-celled plant.

Mantle: Soft fold of integument next to the shell of molluscs.

Meroplankton: Planktonic stages of animals which spend part of their life history in a bottom-living phase.

Metamorphosis: Change in form and structure which takes place when a larva becomes an adult animal.

Moult: Periodic shedding of outer covering. In crustacea followed by increase in size and so essential for growth.

Mucus: Slimy fluid secreted by many animals (adjective 'mucous').

Nauplius: Earliest larval stage of certain crustaceans.

Neritic: Living in coastal waters.

Notochord: Supporting apis along the back of the chordates.

Operculum: Lid or covering flap as in fish's gill cover, the structure closing mouth of shell in some gastropods or movable plates.

Organelles: Various parts of a cell, each with its own special function.

Osculum: Excurrent opening in a sponge.

Osmotic pressure: Pressure created across a semi-permeable membrane when salt solution on one side is more concentrated than on the other.

Pallial line: Mark on inside of bivalve shell where mantle lobes attach. This is indented by the pallial sinus.

Parapodia: Paired, lateral locomotary organs on body segments of polychaetes.

Papillae: Conical eminences on the skin.

Pelagic: Oceanic: inhabiting surface waters or middle depths of the seas.

pH: Degree of acidity or alkalinity. In the scale 0–14, pH7 is neutral. Higher numbers show alkalinity, lower ones acidity. More precisely, the negative power to which 10 is raised to obtain concentration of hydrogen ions in gram-molecules per litre.

Pharynx: Gullet or anterior part of the gut immediately after the mouth.

Photic zone: Surface zone of the ocean penetrated by light.

Photosynthesis: Synthesis of carbohydrate by plants from carbon dioxide and water.

Phyllosoma: Excessively flattened planktonic larval stage of the spiny lobster.

Phytoplankton: Plant plankton.

Plankton: Drifting or floating plants and animals; usually microscopic.

Pneumatophore: The floating air sac of siphonophores.

Poikilotherm: Animal whose temperature varies with that of the surrounding medium; commonly called 'cold-blooded'.

Polyp: A simple member or an individual of a colony of the animal class Anthozoa.

Proboscis: A trunk-like process of the head, as in annelid worms.

Radula: Tongue-like structure bearing chitinous teeth in the mouth of most gastropods.

Raptorial: Predatory or adapted for seizing prey.

Relict species: Animal or plant whose distribution represents the relics of a formerly wider distribution.

Respiration: Gaseous exchange between an organism and its surrounding medium. More specifically oxidation–reduction processes in animals and plants.

Rostrum: Beak-like process projecting between the eyes of certain decapod crustaceans such as the lobster.

Saltmarsh: Area of sand or mud, covered by the sea at high tide and exposed at low tide, inhabited by distinctive flora of salt-tolerant plants.

Scuba: Self-Contained Underwater Breathing Apparatus.

Semi-permeable membrane: Membrane permeable to water but not all dissolved substances.

Sessile: Attached or stationary.

Silica: Mineral substance contained in sponges and diatoms.

Siliceous: Composed of silica.

Siphon: Tube-like organ of bivalve and gastropod molluscs.

Sonar: Equipment that provides echo location in the air or under water.

Spat: Juvenile bivalves, especially oysters.

Spiracle: Respiratory aperture behind the eye in skates and rays.

Spore: Highly specialised reproductive cell of plants.

Sporophyte: Spore-producing phase in alternation of plant generations.

Stamen: Male organ of a flower consisting of a stalk or filament with an anther containing pollen.

Stipe: Stalk; usually in seaweeds.

Stolon: Stem, runner or branch of some animals and plants capable of giving rise to new individuals.

Sublittoral: Below the lowest limits of low tide or the shore.

Substratum: Material upon which a plant grows or on which an animal moves or rests.

Test: Shell or hardened outer covering.

Thallus: Combination of plant cells presenting no differentiation into leaves, stem or roots.

Theca: A structure which serves as protective covering for an organ or a whole organism.

Thorax: Region of an animal behind the head.

Trachea: Windpipe.

Trophic: Pertaining to nutrition or food.

Umbilicus: Basal depression in shells of some gastropods, e.g. top shells.

Umbo: Area behind the beaks in a bivalve shell.

Uropod: Abdominal appendage of a crustacean.

Vascular system: Transport system, e.g. blood in animals or sap in plants.

Velum: Membranous veil-like structure, e.g. ciliated swimming organ of veliger larva.

Ventral: Lower surface of an animal or the surface of a petal or other part which faces centre of flower.

Zoea: Early larval form of certain decapod crustaceans.

Zooid: Individual within a compound animal or colony.

Zooplankton: Animal plankton.

Zygote: Cell formed by the union of two gametes or reproductive cells; fertilised egg.

Bibliography

ANGEL, H. *Photographing Nature—Seashore*. Fountain Press, King's Langley 1975.

BARRETT, J. H. and YONGE, C. M. *Collins Pocket Guide to the Sea Shore*. Collins, 1st ed., London 1958.

BEISER, A. *The Proper Yacht*. Adlard Coles Ltd, London 1978.

BRITISH SUB-AQUA CLUB. *Diving Manual*. British Sub-Aqua Club, 10th ed., London 1977.

BRITISH MUSEUM. *Instructions for Collectors (No. 3): Fishes*. Trustees of British Museum, London 1965.

BRITISH MUSEUM. *Instructions for Collectors (No. 10): Plants*. Trustees of British Museum, London 1957.

BRUNN, B. *The Hamlyn Guide to the Birds of Britain and Europe*. Hamlyn, London 1975.

CAMPBELL, A. C. *The Hamlyn Guide to the Seashore and Shallow Seas of Britain and Europe*. Hamlyn, London 1976.

CHAPMAN, C. F. *Piloting, Seamanship and Small Boat Handling*. Motor Boating and Sailing, New York 1976.

CRAMP, S., BOURNE, W. R. P. and SAUNDERS, D. *The Seabirds of Great Britain and Ireland*. Collins, London 1974.

DOBBS, H. E. *Camera Underwater*. Focal Press, London 1972.

DRUMMOND, MALDWIN. *Conflicts in an Estuary*. Ilex Press, Fawley 1973.

EALES, N. B. *The Littoral Fauna of the British Isles*. Cambridge University Press, Cambridge 1967.

FRIEDRICH, H. *Marine Biology*. Sidgwick & Jackson, London 1969.

GREEN, J. *The Biology of Estuarine Animals*. Sidgwick & Jackson, London 1968.

de HASS, W. and KNORR, F. *The Young Specialist Looks at Marine Life*. Burke, London 1966.

HARDY, A. *The Open Sea: Its Natural History Part I: The World of Plankton*. Collins, London 1956.

HARDY, A. C. *The Open Sea: Its Natural History Part II: Fish and Fisheries*. Collins, London 1959.

HARDY, SIR ALISTER. *Great Waters*. Collins, London 1967.

HEPBURN, I. *Flowers of the Coast*. Collins, London 1952.

HISCOCK, E. C. *Cruising Under Sail*. Oxford University Press, London 1950. (2nd ed. 1965.)

HISCOCK, E. C. *Voyaging Under Sail*. Oxford University Press, London 1959.

KEBLE, MARTIN W. *The Concise British Flora in Colour*. Ebury Press and Michael Joseph, London 1976.

LEWIS, J. R. *The Ecology of Rocky Shores.* The English Universities Press, London 1964.

MEADOWS, P. S. and CAMPBELL, J. I. *An Introduction to Marine Science.* Blackie, Glasgow and London 1978.

MILES, P. M. and H. B. *Seashore Ecology.* Hulton Educational Publications Ltd, Amersham 1966.

MILLAR, R. H. *British Ascidians.* Academic Press, London 1970.

MOOREHEAD, A. *Darwin and the Beagle.* Hamish Hamilton, London 1969.

NEWELL, G. E. and NEWELL, R. C. *Marine Plankton.* Hutchinson Educational, London 1973.

OMMANEY, F. D. *Collecting Sea Shells.* Arco Publications, London 1968.

PRITCHARD, M. (ed.). *The Encyclopedia of Fishing in the British Isles.* Collins, London and Glasgow 1976.

RUSSEL, F. S. and YONGE, C. M. *The Seas.* Frederick Warne, London 1928.

SCHLIEPER, C. *Research Methods in Marine Biology.* V. E. B. Gustav Fischer Verlag Jena, 1968. (Sidgwick & Jackson Biology Series, 1972.)

STANBURY, D. *A Narrative of the Voyage of HMS Beagle.* Folio Society, London 1977.

TEBBLE, N. *British Bivalve Seashells.* British Museum, London 1976.

WELLS, A. L. *The Microscope Made Easy.* Frederick Warne & Co. Ltd, London 1957. (2nd ed. 1969.)

WHEELER, A. *The Fishes of the British Isles and North West Europe.* Macmillan, London 1969.

YONGE, C. M. *The Sea Shore.* Collins, rev. ed., London 1966.

YONGE, C. M. and THOMPSON, T. E. *Living Marine Molluscs.* Collins, London 1976.

Index

219